Bodyspace

Anthropometry, Ergonomics and the Design of Work

Stephen Pheasant
30 March 1949–30 March 1996

Stephen, who died at the tragically early age of 47, will be remembered by a large and diverse group of friends, colleagues, students, courtroom colleagues, and musicians. This alone is testimony to a man whose undoubted intellectual, creative, and communicative skills were matched only by his verve and energy in a wealth of areas.

Stephen was raised in Islington before going up to Gonville and Caius College, Cambridge to read Medical Science, in 1968. His contemporaries will perhaps remember him best for his passion for free jazz and his role in taking the musically based shows 'Stony Ground' and 'Make Me, Make You' to the Edinburgh Fringe in consecutive years. His earlier experience with the National Youth Jazz Orchestra, and the inspiration of his hero Charlie Parker, no doubt influenced him to form the Steve Pheasant Quintet which played at the White Hart Inn, Drury Lane from the mid 1970s to the early 1980s. A close friend and band member, Iain Cameron, recalls Steve's versatility and be-bop creativity on sax, his occasional vocal rendering of 'Let the good times roll' and the band's 'sit in' style, in a manner which reflects the enthusiasm and participative spirit of the man. This, coupled with a burning commitment, are instantly recognized in his professional career.

Students of his at the Royal Free Hospital and University College, where he lectured for many years in anatomy, biomechanics, and ergonomics, could rarely have encountered a more exceptional communicator. His ability to conceptualize and then project complex biomechanical functions in a suitable mode for student learning were testimony to his instinct for education and scholarship. He followed with keen interest the progress of the ergonomists he helped train. His academic and textbook publications were recognized for their application and clarity, a talent acknowledged through the 1985 award, sponsored by the *New Scientist*, for writing about science in plain English. Such skills were inevitably sought by other academic institutions and learned societies, thus he was always high on the invited speaker lists of conference organisers. Professional societies, including the Royal Society of Medicine and the Royal College of Nursing recognized his abilities, as did the British School of Osteopathy, where he held an honorary chair.

His written output was prolific and his textbooks, including the first edition of *Bodyspace* (1986) and *Ergonomics, Work and Health* (1990) have become standards on reading lists around the world. His research output was recognized by the Ergonomics Society with the award of the Sir Frederick Bartlet Medal in 1982, jointly with his close colleague Professor Donald Grieve. His published data of human dimensions have been cited in more ergonomic designs than perhaps any other, and we are grateful too for his contribution to improved design of equipment, tools and many other artefacts of work and leisure use.

When he moved from the academic world, he chose to enter the field of personal injury litigation. In particular, Stephen specialized in work-related musculoskeletal damage, including back pain and repetitive strain injury. As an expert witness, most frequently acting on behalf of the injured party, he was perhaps at his most fulfilled. His desire to challenge orthodoxy, his intellectual skills, his ability to communicate, his love of fierce debate, and his instinct for 'telling a good story' were all given full rein in such an arena. I have rarely seen him happier than when we developed litigious arguments or exchanged courtroom anecdotes with the help of a good Bordeaux. I am sure that adversaries and colleagues alike will sorely miss his presence and his skills.

Stephen knew of his failing health, but never slowed in his endeavours, his output was prodigious. His mother and his partner, Sheila Lee, have much to bear. Family, colleagues, students and friends will remain indebted to Stephen, each in our own way. He will be remembered with affection, respect and regard. I know I speak for many when I say I have lost an inspiring friend.

Dr Peter Buckle

Bodyspace

Anthropometry, Ergonomics and the Design of Work

STEPHEN PHEASANT

SECOND EDITION

TAYLOR & FRANCIS

ALERE FLAMMAM

Founded 1798

UK   Taylor & Francis Ltd, 11 New Fetter Lane, London EC4P 4EE
USA   Taylor & Francis Inc., 325 Chestnut Street, 8th Floor, Philadelphia, PA 19106

British Library Cataloguing in Publication Data

A catalogue record for this book is available from the British Library.

ISBN 0-7484-0067-2 (cased)
ISBN 0-7484-0326-4 (paperback)

Library of Congress Cataloguing Publication Data are available

Cover design by Amanda Barragry

Typeset in Times 10/12 pt by Santype International Ltd, Salisbury, Wiltshire

Printed in Great Britain by TJ International Ltd, Padstow Cornwall

Contents

Foreword

It is now 10 years since the first edition of *Bodyspace* appeared. Over this period of time it has become clear that the science of ergonomics and its application to modern work practices and industrial design have never been needed more. The horrific nature of disasters such as Chernobyl, Bhopal, the Piper Alpha explosion, the Kegworth Air crash and the King's Cross fire have all carried with them important lessons for ergonomists and other designers. The need for an understanding of human behaviour, capacities and needs *prior* to the implementation of a complex system has been identified over and over again. Tragically, the professionals with the required knowledge and skills are too frequently consulted only after the event. I am sure that many of my colleagues would agree that the call to action rarely comes during the design process but rather as a desperate plea following an acute or chronic system failure.

If the major acute complex system failure is the focus of public and media attention then the chronic system failure is the silent enemy. In the UK, a six-fold increase in sickness days lost to back pain since 1974, 1 million workers reporting musculoskeletal problems associated with their work in a single year, and the burgeoning problems of stress-related disorders reflect a society which is neither adapting, managing or designing in sympathy with the needs of the workforce. The cost of this failure is rarely evaluated. The burden of care falls on the tax payer and has been estimated at up to £16 billion.

Organisations – perhaps with some justification – often feel that they are over-regulated and subject to onerous restraints in a highly competitive world. The added 'burden' of health and safety is frequently cited as a limiting factor in the trading success of businesses. I know of no studies which have proven this case and conversely know of many hugely successful organisations who have shown that quality is a broad concept, encompassing issues of product design and production, workforce well-being and environmental impact, amongst others.

It is of concern that the business case for user-focused design is so rarely developed. It is perhaps too obvious that a well-designed tool will perform better in the hands of a skilled operator than a poorly designed one. A failure to document this adequately and regularly leads, too frequently, to good design being replaced by cheaper less effective substitutes. The scope for organisations to improve efficiency by reflecting the goals of ergonomics requires a consideration of the cost of inappropriate work systems, the costs of reduced performance, poor quality, demotivated workforces and ill-health.

Whilst the business case for appropriately designed systems and user designed products has never been stronger, the question remains as to how such goals might be achieved. The knowledge base on which ergonomics rests grows significantly year on year, albeit that the research base, in keeping with most scientific subjects, often raises more questions than it answers. The need for authoritative, contemporary, and usable reference sources is therefore great.

Bodyspace is an example of that rare breed of texts which, upon publication, found favour with both academics and practitioners. Those who knew the author might have anticipated this. It may be twenty years since I first met Stephen but I vividly recall his skills as a lecturer at the Royal Free Hospital and his ability to conjur up a feeling of excitement and a clear understanding of diverse topics in the broad field of human physiology and biomechanics. This feeling was shared by my fellow students and was particularly impressive given that many of those listening were from backgrounds with little prior knowledge of biology.

It has therefore been of no surprise to me, as a Director of a Master's Degree in Ergonomics, that the text which most frequently disappears on a 'permanent loan' from my study has been *Bodyspace*. Indeed, as I write this Foreword I note that the copy in front of me belongs to a colleague!

If the sign of a popular book is its use, then the mark of a good book must be the understanding reached through its content. It is with some relief that, having perused the contents of the new second edition of *Bodyspace*, I note that most of the original valuable material is still there, with the added advantage that the format now reflects a heightened awareness of 'reader usability'.

Of the new material, the chapter on the subject of Health and Safety at Work is most welcome and is most likely to be seen as contentious. The reason for this is the escalating demand for ergonomic expertise in the resolution of litigation between employees and employers following alleged injuries at work. The author's own contributions to this area are reflected here, thankfully without recourse to unergonomic legal 'jargon'.

We will never know how much difference a single text will make to the discipline and application of ergonomics – that *Bodyspace* has come of age with a 2nd Edition is evidence enough that the subject is simultaneously maturing and expanding, whilst continuing to be in increasing demand.

Stephen died on the 30th March 1996, shortly after completing this manuscript. He was acknowledged by his peers as an internationally renowned ergonomist as well as a gifted academic author. *Bodyspace* is testimony to this, and as such, is a significant component of his legacy.

PETER BUCKLE
Reader in Ergonomics and Epidemiology
Robens Institute
University of Surrey

He who would do good to another must do it in Minute Particulars:
General Good is the plea of the scoundrel, hypocrite and flatterer,
For Art & Science cannot exist but in minutely organized Particulars,
And not in generalising Demonstrations of the Rational Power.

William Blake, *Jerusalem*, 1815, pl. 55, ll. 60–64.

I design plain truth for plain people.

John Wesley, Sermon, 1746.

PART ONE

Ergonomics, Anthropometry and the Design of Work

CHAPTER ONE

Introduction

Several similar contests with the petty tyrants and marauders of the country followed, in all of which Theseus was victorious. One of these was called Procrustes or the stretcher. He had an iron bedstead on which he used to tie all travellers who fell into his hands. If they were shorter than the bed he stretched their limbs to make them fit; if they were longer than the bed he lopped off a portion. Theseus served him as he had served others.

From *The Age of Fable* by Thomas Bullfinch (1796–1867)

Prior to her injury, 'Janice' worked as a word processor operator, for a medium-sized firm of management consultants just outside London. She worked in a typing pool with three other girls. One day, one of the partners in the firm needed to get a lot of information entered onto a database in a hurry – and it occurred to him that Janice might work faster if she was in a room on her own where she could not waste time chattering with her friends. So he had a computer terminal set up for her in the firm's library. It was placed on an antique wooden desk. This was somewhat higher than the standard office desk (antiques often are). It had two plinths and a 'kneehole drawer' in the space between them where the user sits. Janice found that however she sat at this desk she could not get into a comfortable working position. She noticed in particular that her wrists were not at their normal angle to the keyboard. It was during the early part of the afternoon that she first began to be aware of a dull ache at the backs of her wrists. This rapidly became worse until she was in considerable discomfort. So she told her boss about it. His response (as it was subsequently alleged) was to say: 'Stop whingeing and get on with your work!' So Janice did. As a result, she developed an acute tenosynovitis affecting the extensor tendons of both wrists. Her condition subsequently became chronic and she was no longer able to type. She lost her job and was forced to take up less well paid employment as a traffic warden. She took legal action against her employers who eventually settled 'on the courtroom steps' for a substantial sum of money.

What lessons may we learn from the story of 'Janice', over and above the more obvious ones concerned with management style and so on? Janice's injury was the result of a *mismatch* between the *demands* of her *working task* and the *capacity* of the muscles and tendons of her forearms to meet those demands. Or to put it another way, the excessive stresses to which these body structures were exposed stemmed from her being forced to *adapt* to an unsatisfactory working position, which was in turn the result of a *mismatch* between the dimensions and characteristics of her *workstation* and those of its *user*.

Injuries of this sort are common enough (although in Janice's case the causative factors in question are perhaps unusually clear-cut ones). Indeed in many parts of the world the incidence of such injuries is said to be reaching epidemic proportions. The problem of musculoskeletal injury at work – important as it may be in both economic and human terms – is but one small facet of a very much larger class of issues concerned with the interactions between human beings and the objects and environments they design and use.

To say that we live in an artificial world is something of a truism. Look around you. It is unlikely that you are reading this in a desert wilderness. More probably you are indoors in a furnished room, or in a moving vehicle, or at least in a culti-vated garden. It is all too easy to ignore the simple fact that most of the visible and tangible characteristics of the artificial environments in which we spend the greater part of our lives are the consequences of design decisions. By no means all of the decisions that lead to the creation of these artificial environments are made by pro-fessional 'designers'. They may be the results of extensive planning or of momentary whims. They represent choices that have been made, which could have been made differently, but were by no means inevitable.

All too often, however, the artefacts that we encounter in our human-made environment are like so many Procrustean beds to which we must adapt. Why should this be so? There is a science that deals with such matters. It is called *ergo-nomics*.

1.1 What is ergonomics?

> Ergonomics is the science of work: of the people who do it and the ways it is done; the tools and equipment they use, the places they work in, and the psychosocial aspects of the working situation.

The word 'ergonomics' comes from the Greek: *ergos*, work; *nomos*, natural law. The word was coined by the late Professor Hywell Murrell, as a result of a meeting of a working party, which was held in Room 1101 of the Admiralty building at Queen Anne's Mansions on 8 July 1949 – at which it was resolved to form a society for 'the study of human beings in their working environment'. The members of this working party came from backgrounds in engineering, medicine and the human sciences. During the course of the war, which had just ended, they had all been involved with research of one sort or another into the efficiency of the fighting man. And they took the view that the sort of research they had been doing could have important applications under peacetime conditions. There did not seem to be a name for what they had been doing, however, so they had to invent one and finally settled on 'ergonomics'.

The word 'work' admits a number of meanings. In a narrow sense it is what we 'do for a living'. Used in this way the activity in question is defined by the context in which it is performed rather than by its content. Unless we have some special reason for being interested in the socioeconomic aspects of work, however, this usage is arbitrary. Some people play the violin, keep bees, bake cakes to make a living; others do it solely for pleasure or for some combination of the two. The content of the activity remains the same.

There is a broader sense, however, in which the term 'work' may be applied to almost any planned or purposeful human activity, particularly if it involves a degree of skill or effort of some sort. In defining ergonomics as a science concerned with human work, we will in general be using the word in this latter and broader sense. Having said this, it would also be true to say that throughout its forty-five years of history, the principal focus of the science of ergonomics has tended to be upon 'work' in the occupational sense of the word.

Work involves the use of tools. Ergonomics is concerned with the design of these – and by extension with the design of artifacts and environments for human use in general. If an object is to be used by human beings, it is presumably to be used in the performance of some purposeful task or activity. Such a task may be regarded as 'work' in the broader sense. Thus to define ergonomics as a science concerned with work, or as a science concerned with design, actually means much the same thing at the end of the day.

The ergonomic approach to design may be summarized in:

The principle of user-centred design
If an object, a system or an environment is intended for human use, then its design should be based upon the physical and mental characteristics of its human users (insomuch as these may be determined by the investigative methods of the empirical sciences).

The object is to achieve the best possible match between the *product* and its *users*, in the context of the (working) *task* that is to be performed (Figure 1.1). In other words:

ergonomics is the science of fitting the job to the worker and the product to the user.

1.1.1 What criteria define a successful match?

The answer to this question will depend upon the circumstances. Criteria that are commonly important include the following:

- functional efficiency (as measured productivity, task performance, etc.);
- ease of use;
- comfort;
- health and safety;
- quality of working life – and so on.

1.1.2 What if these criteria prove incompatible?

Ergonomists often argue that this problem is not as big as it seems. There is some truth in this. There are without doubt circumstances in which ergonomic improvements introduced in the interests of health and safety have a positive pay-off in terms of productivity – and vice versa. Likewise the product that is easy to use will probably, for that very reason, be both safe and efficient in its operation. It is the difficult-to-use products that are, in general, unsafe and inefficient. It would be naive to pretend, however, that these sorts of basic criteria that we have invoked to define a good fit are *never* in conflict, and the deeper we fish in these waters the more difficult the problem becomes.

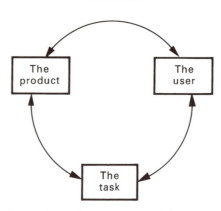

Figure 1.1   User-centred design: the product, the user and the task.

The celebrated American product liability case of *Grimshaw* v. *Ford Motor Company* (1981) is illustrative – notwithstanding that it does not deal with ergonomic issues as such. Briefly the facts were these. The defendants discovered a fault in the design of the petrol tank of one of their models which meant that it was likely to explode in rear-end collisions. On the basis of certain alleged cost–benefit analyses they decided that it would be cheaper in the long run to pay damages for the fatalities and injuries that resulted than to redesign the car and opted to take no further action. Outraged by this cynical view of the economic value of human life and limb, an American jury awarded punitive damages of $125 million against the defendants – which would have gone a long way to wiping out any economic benefits that might have accrued to the defendants from failing to take proper steps to contain the hazard. Regrettably perhaps, this was reduced on appeal to $3.5 million.

This is a somewhat gross example. Cost–benefit trade-offs with implications for health and safety are a fact of everyday industrial life (as any personal injury lawyer will tell you). The fragmented and repetitive short cycle time tasks of the industrial assembly remain an efficient enough way of producing many of the manufactured goods demanded by the consumer economy. But does this production process not have hidden costs? The physical injuries that result are easy enough to recognize – and we could in principle (if we chose) compute the costs of such injuries and incorporate them into some overall system of cost–benefit analysis or social audit. But does it stop there? Does work of this sort result in more subtle sorts of personal injury?

By its very nature, the applied discipline that we call 'ergonomics' sits on the boundary between the domain of empirical science and the domain of ethical values. That is one very good reason that it is important.

1.2   Anthropometrics

Anthropometry is the branch of the human sciences that deals with body measurements: particularly with measurements of body size, shape, strength and working capacity. Anthropometrics is a very important branch of ergonomics. It stands alongside (for example) cognitive ergonomics (which deals with information processing), environmental ergonomics, and a variety of other identifiable sub-disciplines which progress (in parallel, as it were) towards the same overall end.

This book is principally concerned with the anthropometric side of ergonomics: that is, with matching the physical form and dimensions of the product or workspace to those of its user; and likewise with matching the physical demands of the working task to the capacities of the workforce. We shall be developing these issues at length in due course – but first a brief digression.

1.3 Human proportion: an historical perspective

In discussing classical styles of architecture people often use the expression 'designed to the human scale'. The implication is that such buildings are aesthetically well proportioned and convey a certain sense of rightness and harmony. What does this mean? The idea certainly goes back a very long way; and it is closely linked historically to the various 'canons of human proportion' which have been employed by artists and sculptors since ancient times.

The tomb painters of Ancient Egypt (who worked in elevation only and knew no perspective) are known to have employed a modular grid for the preparation of their preliminary drawings of the human figure. The standing figure was divided into fourteen equal parts and the grid intersections corresponded to certain predetermined anatomical landmarks.

Modular systems of this sort (and their equivalents in terms of mathematical ratios between the dimensions of body parts) evolved initially as simple aids to drawing – and indeed rules of thumb of this sort are still taught in life classes today.

In classical times however, the theory of human proportions began to assume a deeper significance; and it came to be thought that certain whole-number ratios between the dimensions of the body and its component parts were inherently 'harmonious' in the sense of being aesthetically pleasing. The argument was probably made, in the first instance, by analogy with musical harmony. The physics of vibrating pipes and stretched strings was known to Pythagoras (*c.* 582–500 BC).

Unfortunately the systems of human proportions used by the sculptors of classical antiquity are for the most part lost to us. The single remnant of these systems which has been passed down to modern times concerns the female nude, in which the nipples and umbilicus are represented as making an equilateral triangle. (We have to allow for the effects of side bending of the trunk, or *contrapposto* as it is called by artists.) You can see this relationship clearly in the Venus de Milo for example, as well as in paintings of the Renaissance and Baroque, as diverse in the actual physical types they portray as Botticelli and Rubens. It is absent, however, in painters who derive their style from Northern Gothic tradition, for example Cranach.

The most detailed system of human proportions which has come down to us from classical times is that of the Roman architectural theorist Vitruvius, writing some time around the year 15 BC. Many of Vitruvius's body-part ratios are familiar to us from archaic units of measurement. The stature of a 'well-made man', for example, is held to be equal to his arm span (one fathom or two yards), which in turn is equal to four cubits (from the elbow to the fingertip), six foot lengths and so on. Vitruvius makes it clear that he regards this 'science' of human proportions as being a fundamental principle in building design.

The celebrated drawing of 'Vitruvian Man' by Leonardo da Vinci, in which a male figure is drawn circumscribed within a square and a circle, must be one of the most overworked visual images around. By Leonardo's day the theory of human proportions had become bound up with that of the so-called 'golden proportion' or 'golden ratio'. It became accepted as a 'fact' that the umbilicus divides the stature of the standing (male) person in golden section: that is, such that the ratio of the greater part to the whole is equal to that of the lesser part of the greater part.

By this stage the entire affair was acquiring distinctly metaphysical overtones. It is these overtones that are invoked perhaps in the expression 'designing to the human scale'. If the phrase has any more pragmatic meaning I have been unable to discern it.

We may think of Leonardo (1452–1519) and his younger contemporary Albrecht Dürer (1471–1528) as standing on the watershed between modern empiricism and the earlier classical tradition, with Leonardo looking backwards and Dürer looking forwards. The classical tradition was prescriptive. It dealt with idealized human beings as they ought to be according to some pre-existing aesthetic or metaphysical principle, rather than real human beings as they actually are. Dürer's *Four Books of Human Proportions*, by contrast, may be regarded as the beginnings of modern scientific anthropometry. In them, Dürer attempts to categorize and catalogue the diversity of human physical types – and his exquisite illustrations are, by his account of the matter at least, based upon the systematic observation and measurement of large numbers of people.

There is a curious footnote to this history. The classical tradition briefly reasserted itself in the middle years of this century in the work of the celebrated French architect Le Corbusier (1887–1965). His definitive treatment of the subject, *The Modulor: A Harmonious Measure to the Human Scale Universally Applicable to Architecture and Mechanics*, is an obscure work thought by many to be profound. It was the same Le Corbusier who said: 'A house is a machine for living in', and thus became one of the patron saints of the school of design known as 'functionalism' (of which more anon).

1.4 Ergonomics and design

What do we mean when we say that a product is 'ergonomically designed'? Regrettably, the short answer to this question is all too often 'not very much'. Nowadays the term is widely used (or misused) in advertising circles. One frequently sees it employed, for example, in the marketing of fancy, overpriced and overdesigned furniture (particularly office furniture) – which is supposed to be 'good for you' in terms of some theory or another (which may or may.not be correct) concerning how to sit correctly. The worst examples I have come across are very expensive indeed and ergonomically quite unsatisfactory. We can of course choose to shrug this off; 'if people are daft enough to buy this, it's their own silly fault'. However, to the responsible professional ergonomist this state of affairs is regrettable in the extreme, not least in that it can only serve to bring his or her profession into disrepute. (We shall return to the ergonomics of furniture in general and office furniture in particular in a later chapter.)

Occasionally, the misuses of the term 'ergonomically designed' have an appealingly surrealist quality. I once came across an account in a Sunday newspaper of 'ergonomically designed pasta', which was (we were told) designed for ease of straining and sauce retention. (I call this fitting the noodles to the user.)

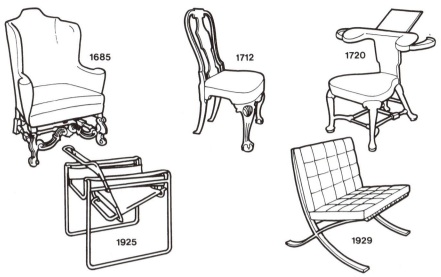

Figure 1.2 Form and function: eighteenth-century style and twentieth-century style. (Upper row: left to right) William and Mary winged armchair, Queen Anne dining chair, early Georgian library chair. (Lower row) 'Wassily' chair by Marcel Breuer, 'Barcelona' chair by Mies van der Rohe. For a contrast see also Figure 4.10.

Here is a good, straightforward, common sense way to recognize an ergonomically designed product, which is quoted from a pamphlet published by the Ergonomics Society (now unavailable) entitled *Ergonomics: Fit for Human Use*.

> Try using it. Think forward to all of the ways and circumstances in which you might use it. Does it fit your body size or could it be better? Can you see and hear all you need to see and hear? Is it hard to make it go wrong? Is it comfortable to use all the time (or only to start with)? Is it easy and convenient to use (or could it be improved)? Is it easy to learn to use? Are the instructions clear? Is it easy to clean and maintain? Do you feel relaxed after a period of use? If the answer to all of these is 'yes' then the product has probably been thought about with the user in mind.

Let us now look a little harder at the issue of functional design – first from the standpoint of design history. The American architect Louis Sullivan is credited with originating the slogan 'form follows function' (*c.* 1895) – his implication being that functional considerations alone are sufficient to determine the form of an object and that ornament is therefore superfluous. According to this theory, functional objects are, of necessity, aesthetically pleasing. This is called 'functionalism'. It was the dominant theory underlying the so-called 'Modern Movement' in design.

When we consider such modern classics as the Marcel Breuer 'Wassily' chair (1925) or the Mies van der Rohe 'Barcelona' chair (1929) we find very little relationship between the form of these seats and that of the human body which it is (presumably) their function to support. The fact that such pieces are commonly referred to as 'occasional chairs' implies that they are without particular function – except to be used 'occasionally'. (In fairness we must admit that the Barcelona chair was in fact designed for the King of Spain to sit on at the opening of an exhibition.)

If we look back to earlier periods of furniture design, for example to the early years of the eighteenth century in Britain, we find a very different state of affairs.

The William and Mary, Queen Anne and early Georgian periods produced furniture in general, and chairs in particular, that showed a closeness of functional
relationship with the human body which has never been excelled (Figure 1.2). Consider the William and Mary winged chair and the variety of ways in which it may
provide the postural support necessary for relaxation; or the Queen Anne dining
chair (sometimes known as the Hogarth chair) with its gently curved back which
reflects the form of the human spine. Neither should we ignore those furniture types
of the Georgian period designed for various very specific functions indeed – the
library or 'cock-fighting' chairs which gentlemen would sit astraddle, the feminine
equivalent for kneeling upon, the reading stands, and even the 'night table' on which
to empty the contents of the pockets. All these bespeak a paramount concern for
user requirements – a relationship between maker and user which is also apparent
in much vernacular design (perhaps most clearly so in the hand tools used by wood-
workers and other craftsmen).

At some time around the midpoint of the eighteenth century, we see function
gradually playing an increasingly accessory role as design was dominated by a suc-
cession of aesthetic theories or 'styles' – neoclassicism, Gothic, etc. Paradoxically,
the most recent of these styles is called 'functionalism', but it should be seen as an
aesthetic demand for absence of ornament, 'truth to materials', etc., rather than a
particular concern with end use. Functionalism is essentially a visual metaphor by
which a designed object may acquire certain desirable connotations.

There was a period some years ago when I used to spend a fair amount of my
time teaching design students. The discussions that we had were the origin of the
five fundamental fallacies set out in Table 1.1. They revolve around two principal
themes. The first is the contrast between the investigative methods of the empirical
sciences and the creative problem-solving methods of the designer which, for want
of a better word, we could call 'intuitive'. The second theme is that of human diversity. In my view this is the single most important characteristic of people to be
borne in mind in the world of practical affairs in general and of design in particular.
To put it plainly, people come in a variety of shapes and sizes – to say nothing of
their variability in strength, dexterity, mentality, and taste. As we shall see, the five
fallacies are increasingly difficult to refute.

Not many people would express the first fallacy in so many words, but in implicit
form it is very widespread. How many products are actually tested at the design

Table 1.1 The five fundamental fallacies.

No. 1	This design is satisfactory for me – it will, therefore, be satisfactory for everybody else.
No. 2	This design is satisfactory for the average person – it will, therefore, be satisfactory for everybody else.
No. 3	The variability of human beings is so great that it cannot possibly be catered for in any design – but since people are wonderfully adaptable it doesn't matter anyway.
No. 4	Ergonomics is expensive and since products are actually purchased on appearance and styling, ergonomic considerations may conveniently be ignored.
No. 5	Ergonomics is an excellent idea. I always design things with ergonomics in mind – but I do it intuitively and rely on my common sense so I don't need tables of data or empirical studies.

stage on a representative sample of users? More commonly the evaluation of a design proposal is entirely subjective. The designer considers the matter, tries out the prototype and concludes that it 'feels alright to me', with the clear implication that if it is satisfactory 'for me' it will be for other people too. In general, objects designed by the stronger or more able members of the population can create insurmountable difficulties for the weaker and less able. Women frequently say with exasperation, 'You can tell it was designed by a man!'

The first fallacy is closely linked with the last by the concept of empathy, of which more anon; it is also closely linked to the second since most people consider themselves to be more or less average. Suppose we were to determine the dimensions of a door by the average height and breadth of the people who were to pass through it. The 50% of people taller than average would bang their heads; the 50% wider than average would have to turn sideways to squeeze themselves through. Since the taller half of the populations are not necessarily the wider half, we would, in fact, satisfy or accommodate less than half of our users. Nobody would make such an elementary mistake in designing a door – but, in my experience, the second fallacy turns up quite frequently in the work of students, of both design and ergonomics, who have only partially grasped the principles of anthropometrics. Obviously enough, we must seek to accommodate the largest percentage possible of the user population (see Chapter 2).

The third fallacy really has the ring of truth. Human beings are indeed very adaptable – they will put up with a great deal and might not necessarily complain. In the example we have just quoted, the taller half of the population would presumably learn to duck. This is the Procrustean approach to design. Adaptation to the Procrustean bed commonly has 'hidden costs' in terms of ill health, although only rarely are these as dramatic as an amputated limb. Consider the economic losses occasioned by the extensive range of musculoskeletal disorders which may be attributed to faulty workspace design – back pain, neck pain, repetitive strain injuries, and so on (see Chapter 8).

Part of the refutation of the third fallacy rests upon these hidden costs of adaptation. In addition we should consider that the design process not only responds to consumer needs but in some measure creates them as well. We could question the extent to which (a) the public gets what the public wants; (b) the public wants what the public gets; or (c) the public knows perfectly well what it wants, but can't get it and puts up with whatever is available. Superimposed over these possibilities are the effects of marketing and advertising on the one hand and consumer pressure groups and legislation on the other. The objects that the designer creates reflect the society in which they are created. In some cases, consumer pressure leads to the introduction of ergonomic features into design. This has happened quite dramatically in recent years in the area of office technology. The computer workstations of today are very much better than those of a decade or more ago – principally because of the effects that consumer pressure has had on market forces. In some areas, consumers are prepared to pay extra for quality. In a later section we shall consider the desirability of providing kitchen worksurfaces at a range of heights – this is perfectly possible technically but is generally deemed uneconomic. For which 'quality' would the informed consumer rather pay extra – an elegant finish with gleaming worktops and polished brass door furniture, or ease of use and less backache? However, beyond all these considerations is the simple fact that making something the right size is often no more expensive than making it the wrong size. The decision to ignore ergonomics on grounds of economics is often just an excuse.

The fifth and final fallacy involves some rather complex issues. The intuition and common sense of which we speak in this context is sometimes called 'empathy' – and if you are a designer you may well have it in abundance. (Whether it is an innate gift or the fruit of experience is another matter.) Empathy is an act of introspection or imagination by which we may 'place ourself in another person's shoes'. It could be argued that, by empathetically casting oneself in the role of the user, the act of designing for others becomes an extension of designing for oneself and the traditional subjective approach becomes valid. In some measure this is probably true, but can these intuitions really circumvent the problems of human diversity? Can we really imagine how somebody quite different from ourselves would experience a certain situation?

As far as I am aware this question has never really been put to the test. Psychologically it is a very interesting one. In general we would predict that empathy would increase with things like social and demographic proximity (as measured by age, sex, etc.); or with similarity in physical characteristics such as strength and fitness, attitudinal characteristics and so on. For any given degree of proximity or similarity, we should obviously expect some people to be more empathic than others. Were we able to measure this trait we might well find that it correlates in interesting ways with other personality characteristics. What sorts of people are the most empathic? Regrettably, however, this all remains within the realms of speculation.

The term 'common sense' also deserves some scrutiny, not least because you often hear people say (perhaps with a measure of truth), 'Ergonomics – that's just common sense!' As a rule, statements like this should be viewed with circumspection. At one time the term *sensa communis* was used to refer to a (hypothetical) physiological system which integrated the separate functions of the traditional 'five senses' of vision, hearing, touch, taste and smell. 'Common sense' underwent a major shift of meaning, however, with its modern usage (as far as I am able to tell) being established by the eighteenth century or thereabouts. We all think we know what it means because we all have it. At one level expressions like 'that's just common sense!' can be used as a justification for the blind acceptance of an untested hypothesis. We must also distinguish 'common sense' from 'common knowledge' and the 'conventional wisdom'. There are those who think that common sense and the scientific method are much the same thing – the latter being a refined version of the former. It seems to me that there is a good deal of truth in this. I would only add that common sense sometimes seems remarkably rare.

1.4.1   The user-centred approach

We have described the ergonomic approach to design as user-centred. How may we characterize this description more fully? One way would be in terms of methodology. In the forty-five years of its history, the science of ergonomics has built up both a substantial organized body of knowledge about human capacities and limitations, and a repertoire of investigative methods for acquiring such knowledge and for practical problem-solving. Two particular techniques deserve special attention: *task analysis* and the *user trial*. In the days when I taught design students I often used to say 'Every good project starts with a task analysis and ends with a user trial'. But they never showed much sign of taking notice – much preferring their boxes of magic markers. Task analysis and user trials are both extremely simple in concept – to the point perhaps of being 'just common sense'.

A *task analysis* is really a formal or semi-formal attempt to define and state what the user/operator is *actually going to do* with the product/system/environment in question. This is stated in terms of the desired ends of the task, the physical operations the user will perform, the information-processing requirements it entails, the environmental constraints that might pertain, and so on. An effective task analysis will clarify the overall goals of the project, establish the criteria that need to be met, point out the most likely areas of *mismatch*, and so on.

A *user trial* is just what its name suggests: an experimental investigation in which a sample of people test a prototype version of the product under controlled conditions. The subjects in the trial must be chosen with care. Ideally they should be a representative sample of the population of users for whom the end product is ultimately intended. There would be little point in trying out some new high-tech product on the technophiles down the corridor if it is ultimately going to be used by

---

Table 1.2   User-centred design.

1. *User-centred design is empirical*
   It seeks to base the decisions of the design process upon hard data concerning the physical and mental characteristics of human beings, their observed behaviour and their reported experiences. It is distrustful both of grand theories and intuitive judgements – except insomuch as these may be used as the starting points for empirical studies.

2. *User-centred design is iterative*
   It is a cyclic process in which a research phase of empirical studies is followed by a design phase, in which solutions are generated which can in turn be evaluated empirically.

3. *User-centred design is participative*
   It seeks to enrol the end-user of the product as an active participant in the design process.

4. *User-centred design is non-Procrustean*
   It deals with people as they are rather than as they might be; it aims to fit the product to the user rather than vice versa.

5. *User-centred design takes due account of human diversity*
   It aims to achieve the best possible match for the greatest possible number of people.

6. *User-centred design takes due account of the user's task*
   It recognizes that the match between product and user is commonly task-specific.

7. *User-centred design is systems-orientated*
   It recognizes that the interaction between product and user takes place in the context of a bigger socio-technical system, which in turn operates within the context of economic and political systems, environmental ecosystems, and so on.

8. *User-centred design is pragmatic*
   It recognizes that there may be limits to what is reasonably practicable in any particular case and seeks to reach the best possible outcome within the constraints imposed by these limits.

the technophobes in the street. Sometimes, as a deliberate strategy, it makes sense to test a product on those sorts of people who are likely to have most difficulty in using it – the technologically naive, the elderly and infirm, and so on – on the grounds that if they can cope, then the product will also be acceptable for the more able majority. (This is the equivalent of the *principle of the limiting user* in anthropometrics which we shall encounter in the next chapter.) We must likewise take care to ensure that the circumstances under which the trial is conducted are a reasonably valid approximation to those of real-world use.

For a more detailed treatment both of task analysis and user trials and of ergonomics methodology in general, the reader is recommended to turn to Wilson and Corlett (1995).

An alternative way of characterizing the user-centred approach might be in terms of its recurrent features. I have attempted to summarize what I regard as the most important of these in Table 1.2 above. This will also serve as a summary of much of that which has gone before in this chapter.

CHAPTER TWO

Principles and practice of anthropometrics

There are a few situations in which it is possible to design a product or workstation for a single user: bespoke tailoring, *haute couture*, the customized seats used by racing drivers, and the workstations of astronauts are examples. These are all essentially *luxury goods*. For a very small number of especially unfortunate individuals the 'luxury' of custom design becomes a necessity. The physical characteristics of the very severely disabled are so diverse that aids to mobility and independence must often be made for the individual concerned. But in the very great majority of real-world design problems our concern will be with a population of users.

We all acknowledge the necessity of manufacturing garments in a range of sizes, but would it also be true to say that chairs and tables, for example, should be supplied in a range of sizes as well? The answer is 'only to a limited extent'. We do not expect adults and children to use the same-sized writing desks in their offices and schools; although they seem to cope perfectly well with the same dining table at home. We commonly supply typists with adjustable chairs; but their desks are usually of fixed height. Obviously enough, we are prepared to accept a less accurate fit from a table and chair than from a shirt and trousers. What is rather less obvious is how we should choose the best compromise dimensions for equipment to be employed by a range of users, and at what point we should conclude that adjustability is essential. In order to optimize such decisions we require three types of information:

(i) the anthropometric characteristics of the user population;
(ii) the ways in which these characteristics might impose constraints upon the design;
(iii) the criteria that define an effective match between the product and the user.

Before discussing these matters further we shall need to establish some of the mathematical foundations upon which the applied science of anthropometrics rests. In the section that follows I have endeavoured to do this with the minimum possible recourse to the use of mathematical equations and formulae. The reader who requires a more detailed mathematical treatment of the subject is referred to the Appendix.

2.1   The statistical description of human variability

In order to establish the statistical concepts that describe human variability, let us conduct what earlier scientific writers would have called an experiment of the imagination. Supposing you are in a large public building frequented by a fairly typical cross-section of the population. A companion, who is an inveterate gambler, offers to take bets on the stature (standing height) of the next adult man to walk down the corridor. (We could just as well bet on women, children, or everyone taken together – but it is a little easier to deal with the problem mathematically if we only consider adults of one sex.) On what height would you be best advised to place your money (assuming of course that you have no prior knowledge of people who happen to be in the area)? You will probably pick a stature that is somewhere near the average, since experience has told you that middling-sized people are relatively common, whilst tall or short people are rare by comparison. You have in essence made a judgement as to the relative probability of people of different statures, or the relative frequency with which such people are encountered by chance. Average people are more probable than extremes, in that you encounter them more frequently. The statistically minded punter, offered a bet of this kind, could optimize his chance of winning by going out and measuring all the men in the building. With these data we could plot a chart like the one shown in Figure 2.1, in which probability (frequency of encounter) is plotted vertically against stature, horizontally. The smooth curve on this chart is known as a probability density function or a frequency distribution. The particular curve we have drawn here is symmetrical about its highest point –

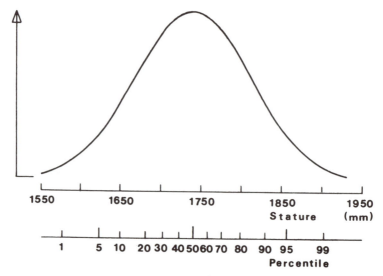

Figure 2.1   The frequency distribution (or probability density function) for the stature of adult British men. This is an example of the normal or Gaussian distribution. (Data from Knight (1984).)

the average stature, otherwise known as the mean, is also the most probable stature. Since the curve is symmetrical, it follows that 50% of the population are shorter than average and 50% are taller – we would say, therefore, that in this distribution the mean is equal to the 50th percentile (50th %ile). In general, n% of people are shorter that the nth %ile. Hence, somewhere near the left-hand end of the horizontal axis there is a point, known as the 5th percentile (5th %ile) of which we could say 'exactly 5% of people are shorter than this' or 'there is only a one-in-twenty chance of encountering a person shorter than this'. Similarly, an equal distance from the mean towards the right of the chart is a point known as the 95th %ile of which we could say 'only 5% of people are taller than this'. Ninety per cent of the population are between the 5th and 95th %ile in stature – but the same could be said for the 2nd and 92nd or the 3rd and 93rd. It is important to note that, by virtue of their symmetrical positions about the mean, the 5th and 95th %ile define the shortest distance along the horizontal axis to enclose 90% of the population. Two further points must be borne in mind when discussing percentiles. Firstly, percentiles are specific to the populations that they describe – hence, the 95th %ile stature for the general public might only be the 70th %ile for a specially selected occupational group like the police force or perhaps the 5th %ile for a sample made up from the Harlem Globetrotters and other professional basketball teams. Secondly, percentiles are specific to the dimension that they describe – hence, a person who is a particular percentile in stature may or may not be the same in shoulder breadth or waist circumference, since people differ in shape as well as in size.

The frequency distribution shown in Figure 2.1, with its characteristic symmetrical bell-shaped curve, is very common in biology in general and anthropometry in particular. It is usually known as the normal distribution. We should not, however, infer from this name that the distribution is in some way associated with 'normal people' as against 'abnormal' ones. We might conveniently think of the term as meaning something like 'the distribution which you will find most useful in practical affairs'. To avoid this possibility of confusion, some statisticians prefer to call it the 'Gaussian distribution' – after the German mathematician and physicist Johann Gauss (1777–1855) who first described it (in the context of random errors in the measurement of physical quantities). It is possible to predict that a variable such as stature will be normally distributed in the general population if we are prepared to make certain plausible assumptions concerning the way it is inherited from one generation to the next (see any textbook of genetics). Indeed, it is empirically true that most anthropometric variables conform quite closely to the normal distribution (at least within reasonably homogeneous populations). This is an exceedingly convenient state of affairs since the normal distribution may be described by a relatively simple mathematical equation. The exact form of this equation need not concern us here since we are unlikely to employ it in practice. The important thing is that it has only two parameters. (In mathematics a parameter is a quantity that is constant in the case considered but variable in different cases.) One of these parameters is the mean – it tells us where the distribution is located on the horizontal axis. The other is a quantity known as the standard deviation (SD) which is an index of the degree of variability in the population concerned, i.e. the 'width' of the distribution or the extent to which individual values are scattered about or deviate from the mean. If we were to compare, for example, the general male population with the police force, we would find that the latter had a greater mean but a smaller standard deviation, i.e. they are on average taller than the rest of us and they are less variable amongst themselves. The standard deviation (SD) of a sample of individuals drawn from a population is given by the equation

$$SD = \sqrt{\frac{\Sigma(x - m)^2}{n - 1}} \tag{2.1}$$

where m is the mean, x is any individual value of the dimension concerned and n is the number of subjects in the sample. (We use $n - 1$ in the equation in the hope of correcting any bias introduced by the finite size of our sample and making a better prediction of the standard deviation of the population from which it was drawn – since this is what in general concerns us.)

A normal distribution is fully defined by its mean and standard deviation – if these are known any percentile may be calculated without further reference to the raw data (i.e. the original measurements of individual people). The pth %ile of a variable is given by

$$X_{(p)} = m + sz \tag{2.2}$$

where z is a constant for the percentile concerned, which we look up in a statistical table.

A selection of z values for some important percentiles are given in Table 2.1. For a more detailed table of p and z, turn to the beginning of the Appendix. Suppose we wish to calculate the 90th %ile of stature for the adult male population of Britain. It happens that British men have a mean stature of 1740 mm with a standard deviation of 70 mm (see Table 4.1). From Table 2.1 we see that for $p = 90$, $z = 1.28$. Therefore the 90th %ile value of stature $= 1740 + 70 \times 1.28 = 1824$ mm. Alternatively, we might wish to do the calculation in reverse, and determine the percentile value for a particular stature. Hence, a stature of 1625 mm is 1.64 standard deviations below the mean. That is $z = -1.64$. Looking this up in Table 2.1 we find that this is equivalent to the 5th %ile.

In this book we shall, in the interests of brevity, commonly adopt a convention for describing the parameters of normal distributions. Whenever a figure is followed by another in square brackets [] it refers to a mean and standard deviation. Hence, the statement that 'the stature of British men is 1740 [70] mm' should be taken as meaning 'the stature of British men is normally distributed, with a mean of 1740 mm and a standard deviation of 70 mm'. (This is a purely local convention; you will not encounter it outside this book.)

Table 2.1 Values of z for selected percentiles (p)

p	z	p	z
1	−2.33	99	2.33
2.5	−1.96	97.5	1.96
5	−1.64	95	1.64
10	−1.28	90	1.28
25	−0.67	75	2.67
50	0.00		
0.1	−3.09	99.9	3.09
0.01	−3.72	99.99	3.72
0.001	−4.26	99.999	4.26

Most linear dimensions of the body are normally distributed and this certainly makes life easier for the user of anthropometric data. There are, however, other kinds of frequency distribution which turn up occasionally in anthropometric practice. Some other possibilities are shown in Figure 2.2. In most populations body weight and muscular strength show a modest positive skew – it seems that there are a disproportionate number of heavy, strong people and a dearth of light, weak ones. Furthermore, the combination of two normal distributions, such as men and women or adults and children, will give us a new distribution that is flat-topped (platykurtic) or even double-peaked (bimodal). What will happen if we work on the erroneous assumption that such distributions are normal and go ahead and calculate percentiles by the means described above? Errors will accrue, the magnitude of which will be determined by the extent of the deviation from normality in the population distribution. The errors will in many circumstances be negligible. Combining data for adult men and women is a case in point. In theory, the resultant 'unisex' distribution is platykurtic. In practice, the deviations from normality are so small that we can ignore them. The only alternative, which avoids the assumption of normality, is to determine percentiles directly by simply counting heads – but since this requires large numbers of subjects it is rarely feasible and few datasets in the literature have been established with this degree of certainty. In general, the best practice is to assume normality but to proceed with circumspection in those situations (mentioned above) where we have reason to doubt the assumption. From now on, our discussion will be almost entirely confined to normal distributions.

For some purposes it may be especially informative to plot out the normal distribution in its cumulative (or integral) form. In this version percentiles are plotted against values of the dimension concerned (or, if we calibrate the horizontal axis in standard deviations, we have in effect a plot of p against z). The curve that we obtain is known as the normal ogive – see Figure 2.3. The advantage of such a plot is that, since we may read off percentiles directly, it enables us to evaluate the consequences of a design decision for the percentage of users accommodated. To take a simplistic example, Figure 2.3 would tell us directly the percentage of British men who could pass beneath an obstruction of a given height without stooping or banging their heads.

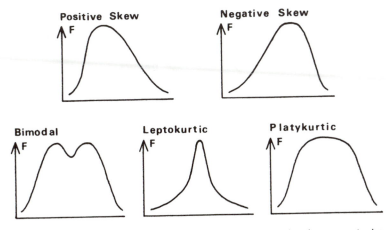

Figure 2.2 Deviations from normality in the statistical distributions of anthropometric data.

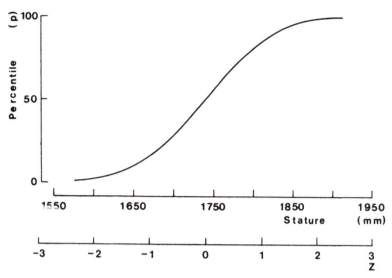

Figure 2.3   The frequency distribution of the stature of adult British men, plotted in cumulative form.

The slope of the normal ogive is greatest at the mean value (i.e. the point of maximum probability) and steadily diminishes as we approach the extreme tails of the distribution. The curve is asymptotic to the horizontal at 0 and 100% (i.e. in theory meets these lines at infinity). Hence, it is increasingly difficult to accommodate extreme percentiles. (We note in Figure 2.1 that the percentiles are densely packed near the centre and thinly spread at the extremes.) The practical consequence of this is that each successive percentage of the population we wish to accommodate imposes a more severe constraint upon our design. In cost/benefit terms we are in a condition of steadily diminishing returns. Figure 2.4 illustrates this problem with respect to the case of the adjustability of a seat. Calculations were

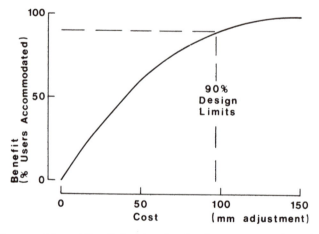

Figure 2.4   Anthropometric cost/benefit function showing the percentage of members of a target population accommodated by various ranges of adjustment in the height of a seat. The range plotted should in each case be 'centred' on a height of 455 mm.

based on the criterion that seat height should be equal to popliteal height, which for the unisex distribution of adult British men and women (shod) = 455 [30] mm.

How then should we draw the line in this increasingly costly and constraining process of accommodating the extreme members of the user population? In other words, where should we set our *design limits*? A purely arbitrary answer to this question, which has been found to work well enough in practice for many purposes, is to design for the 5–95 %ile range, i.e. for the middle 90% of the user population. When using this rule of thumb, however, we must always bear in mind the consequences of a *mismatch* for those 10% of the user population who are outside our design limits. Will a mismatch merely cause mild discomfort and inconvenience? Or might it compromise the overall working efficiency of the system? Are there implications for the health and safety of the user, in either the short or long term? A less than 5th %ile person sitting at a dining chair that is too high may be mildly uncomfortable over dinner; but if she has to work at a desk that is too high for seven hours a day, five days a week, the consequences may be very much more severe (see Chapters 7 and 8).

Supposing we were asked to specify dimensions for an escape hatch from a confined working area. A value based upon the appropriate bodily dimensions of a 99th %ile user would mean that one person in 100 would get stuck. This would clearly not be acceptable. (Actually the problem is compounded by the fact that the distributions of the body bulk dimensions involved will probably be positively skewed.) In safety-critical applications of this sort, each individual case must be judged on its own merits. We might, for example, decide, as a matter of policy, that less than one user in 10 000 should be mismatched – and set our design limits at around four standard deviations from the mean (see Table 2.1).

In a more general sense, it is only possible to specify percentiles at all if we can first define the user population. This would be simple enough in the case of a fighter aircraft, for example – but the users of a public transport system would be quite another matter. Must we consider children as well as adults – or the elderly and infirm, pregnant women or wheelchair users? These people may not fit readily into the percentile tables of the anthropometrist; but can they legitimately be excluded from participation in the system or environment in question? We shall return to the issue of *barrier-free design* in due course, but first we must deal with the narrower problem of designing for the majority.

2.2 Constraints and criteria

In ergonomics and anthropometrics a constraint is an observable, preferably measurable, characteristic of human beings, which has consequences for the design of a particular artefact. A criterion is a standard of judgement against which the match between user and artefact may be measured. We may distinguish various hierarchic levels of criteria. Near the top are overall desiderata such as comfort, safety, efficiency, aesthetics, etc., which we may call high-level, general, or primary criteria. In order to achieve these goals, numerous low-level, special, or secondary criteria must be satisfied. The relationship between these concepts may be illustrated by way of example. In the design of a chair, comfort would be an obvious primary criterion; the lower leg length of the user imposes a constraint upon the design since, if the chair is too high, pressure on the underside of the thigh will cause discomfort. This leads us to propose a secondary criterion: that the seat height must not be greater than the vertical distance from the sole of the foot to the crook of the knee (this dimension is called popliteal height). A table of data will tell us the distribution of this dimension. It would seem reasonable to choose the 5th %ile value – since if a person this short in the leg is accommodated, the 95% of the population who are longer legged will also be accommodated. This leads more or less directly to a design specification or tertiary criterion: that the height of the seat shall not be greater than 400 mm. (Note that if we propose an adjustable seat we will use our criterion differently, as in Figure 2.4 – see Chapter 4 for a more general discussion of this particular problem.)

   Taken in isolation, the primary criterion will usually be what is known, amongst certain ergonomists of my acquaintance, as a 'stunning glimpse of the obvious' (SGO). In general, it is necessary to work down through successive levels of the hierarchy before any operationally useful recommendations result. Some theorists like to contrast the 'top down' approach of working from the general to the specific with the 'bottom up' approach of working from the specific to the general.

   At any level in the hierarchy conflicts between criteria may arise which will necessitate trade-offs. Hence, in the example we took above, our secondary criterion tells us when a seat is too high but not when it is too low. The criteria for this latter case are less well defined – we might call them fuzzy rather than sharp. None the less, it is perfectly possible that a tall man might feel uncomfortably cramped in a seat designed to accommodate the lower leg length of a 5th %ile woman, and some suitable compromise might have to be reached in the interest of the greatest comfort for the greatest number. Similarly, there might be circumstances in which it was necessary to trade-off, say, comfort against efficiency or safety. I suspect that these latter circumstances are few, but they raise the interesting point of what super-ordinate criterion could be used to measure both.

   In practical matters, the middle of the hierarchy is often the best place to start (I have heard this called the 'middle-out' approach). We shall therefore consider four sets of constraints which between them account for the vast majority of everyday problems in anthropometrics *per se* and, hence, a sizeable portion of ergonomics. We shall call them the four cardinal constraints of anthropometrics: clearance, reach, posture, and strength.

2.2.1 Clearance

In designing workstations it is necessary to provide adequate head room, elbow room, leg room, etc. Environments must provide adequate access and circulation space. Handles must provide adequate apertures for the fingers or palm. These are all clearance constraints. They are one-way constraints and determine the minimum acceptable dimension in the object. If such a dimension is chosen to accommodate a bulky member of the user population (e.g. 95th %ile in height, breadth, etc.), the remainder of the population, smaller than this, will necessarily be accommodated.

2.2.2 Reach

The ability to grasp and operate controls is an obvious example – as is the constraint mentioned above on the height of a seat or the ability to see over a visual obstruction. Reach constraints determine the maximum acceptable dimension of the object. They are again one-way constraints, but this time are determined by a small member of the population, e.g. 5th %ile.

2.2.3   The principle of the limiting user

The limiting user is that hypothetical member of the user population who, by virtue of his or her physical (or mental) characteristics, imposes the most severe constraint on the design of the artefact. In clearance problems the bulky person is the limiting user; in reach problems the small person is the limiting user.

2.2.4   Posture

A person's working posture will be determined (at least in part) by the relationship between the dimensions of his body and those of his workstation. Postural problems are commonly more complex than problems of clearance and reach – since we may well have limiting users in both tails of the distribution. For example, a working surface that is too high for a small person is just as undesirable as one that is too low for a tall person (see below, and also section 3.6). In other words we have a two-way constraint.

2.2.5   Strength

A fourth constraint concerns the application of force in the operation of controls and in other physical tasks. Often, limitations of strength impose a one-way constraint, and it is sufficient to determine the level of force that is acceptable to a weak limiting user. There are cases, however, where this may have undesirable consequences for the heavy-handed (or heavy-footed) user, or in terms of the accidental operation of a control, etc. – and in these cases a two-way constraint may apply.

2.3   Fitting trials and the method of limits

A fitting trial is an experimental study in which a sample of subjects use an adjustable mock-up of a workstation in order to make judgements as to whether a particular dimension is 'too big', 'too small', or 'just right'.

   Figure 2.5 shows the results of a simple fitting trial, the purpose of which was to determine the optimum height for a lectern in a lecture theatre. Ten students (five male, five female) acted as subjects. A music stand served as the adjustable mock-up. Each subject set the music stand to the highest and lowest heights that he or she considered acceptable – and then to their own personally preferred optimum height. The mean and the standard deviations of the upper and lower limits for all ten subjects were calculated. These were used to plot the smooth curves for 'too low' and 'too high' shown in Figure 2.5 (by using z and p values as described above). The smooth curve for 'satisfactory' is plotted by calculating the percentage of people for whom it was neither 'too low' nor 'too high'. (Thus for any given height, 'too low' + 'too high' + 'satisfactory' = 100%.) The distribution for 'just right' was

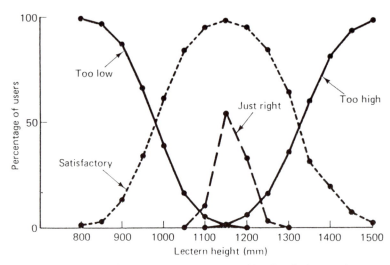

Figure 2.5 Results of a fitting trial to determine the optimal height of a lecturn. (From A. S. Nicholson and J. E. Ridd, *Health, Safety and Ergonomics*, Butterworth-Heinemann, 1988, Figure. 7.2; reproduced with kind permission.)

derived by rounding off each individual subject's response to the nearest 50 mm and plotting these directly.

In this experiment, a clear optimum of 1150 mm emerged (at least for this group of subjects). At this height, more than 50% of users regard the lectern as 'just right' (within ± 25 mm), and over 95% consider it satisfactory.

The design of the experiment could perhaps have been improved. It might, for example, have been better to start with a very high lectern and then come down in 50 mm steps, with each subject making a rating at each height, until a point was reached at which the lectern was definitely too low. Then to repeat the process going up, until a point was reached at which it was definitely too high. (Curiously enough, when you do it like this the ascending and descending trials tend to give you slightly different answers. The descending trial generally gives a lower optimum.) We could also question the extent to which the subjects in this experiment were a representative sample of the 'true' population of end-users of the lectern. If the problem were more critical it might merit a more detailed investigation. But for many real-world questions a relatively crude experiment of this nature will suffice.

A fitting trial is a type of *psychophysical experiment* – that is, one in which people make *subjective* (i.e. psychological) judgements concerning the *objective* properties of physical objects or events. The form that the plotted results of the lectern experiment took is very characteristic. It is made up from the ogival curves of two normal distributions, facing each other, which define a third normal distribution by substraction of their summed values from 100%. Similar results may be found in other areas of ergonomics. Judgements about thermal comfort, for example – 'too hot', 'too cold', 'just right' – are distributed in this way (see Grandjean 1986). In principle we might expect to encounter this form in any situation in which people are asked to express a *subjective preference* on a bipolar *continuum*: too fat/too thin, too young/too old, and so on.

Consider now the problem of seat height which we discussed in Section 2.2. We could in principle have solved this by conducting a fitting trial – but instead we solved it by the application of anthropometric data. The line of reasoning that we adopted could be written out in a formal way as follows:

- criterion: seat height ⩽ popliteal height;
- limiting user: 5th %ile woman, shod popliteal height = 400 mm;
- design specification: maximum seat height = 400 mm.

Let us now apply a similar line of reasoning to the analysis of a more complex problem involving a two-way postural constraint. The technique we are going to use is called the *method of limits*. (The name is borrowed from that of a technique in psychophysics which is equivalent in form.) In essence this technique is a model or analogue of the fitting trial, in which anthropometric criteria and data are used as substitutes to 'stand for' the subjective judgements of real people.

The problem is to determine the optimum height for a workbench to be used in a certain industrial task which involves a moderate degree of both force and precision. To simplify the calculation we shall assume that the task will be performed by male workers. The workers will be standing. According to Grandjean (1986) the optimum working height for a task involving moderate force and precision is between 50 and 100 mm below the person's elbow height. We note that it is a two-way criterion since it may be exceeded in either direction. The elbow height (EH) of British men is 1090 [52] mm (see Table 2.3). To this we must add a 25 mm correction for shoes giving 1115 [52] mm (see Section 2.4). Combining these data with the above criterion gives us the upper and lower limits of optimal working level. EH − 50 = 1065 [52]; EH − 100 = 1015 [52]. We can treat these just as if they are new normally distributed anthropometric dimensions – and calculate the percentile in these distributions to which any particular workbench height corresponds. However, we

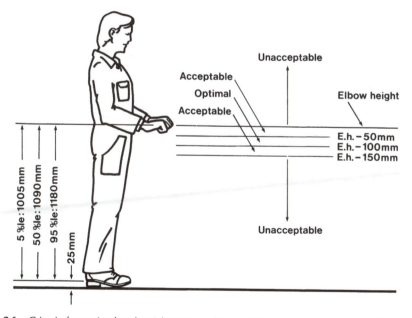

Figure 2.6 Criteria for optimal and satisfactory working heights in an industrial assembly task.

Table 2.2   Calculation of percentage of men accommodated by a workbench that is 1000 mm in height

Criterion	Distribution	Percentile	Conclusion
EH-150	965 [52]	75	25% – much too low
EH-100	1015 [52]	39	61% – too low
EH-50	1065 [52]	11	11% – too high
EH	1115 [52]	1	1% – much too high
			28% – just right

should bear in mind that the criteria refer to 'optimal' bench heights. Since we may
reasonably assume that users may be prepared to accept less than absolute per-
fection, we may well find it useful to consider two further zones extending 50 mm
above and below the optimum, which we would characterize as 'satisfactory but not
perfect' (see Figure 2.6). We choose the figure of 50 mm because it 'seems reasonable'
rather than on the basis of any particular scientific evidence.

Table 2.2 shows a complete set of calculations performed for the height of 1000
mm. We find that this corresponds to the 75th %ile in the lowest criterion distribution – from which we infer that a workbench of 1000 mm would be 'much too low',
or 'unsatisfactory', for the 25% of men who are taller than this. Similarly, the centre
criteria correspond to the 39th and 11th %iles, respectively – from which we infer
that the 28% of men between these heights would find the workbench 'just right' or
'optimal'.

We could keep on performing such calculations for different heights until we
homed in on a value that maximized the percentage optimally matched and minimized the unsatisfactory matches. (Here, of course, the computer would help.) At
this point we are like the statistically minded punter searching for the best bet. The
results of a series of such calculations are plotted in Figure 2.7. It comes as no
surprise to discover that the 'optimal' figures describe a normal curve (e), whereas
the 'too high' and 'too low' figures yield normal ogives facing in opposite directions
(a, b, c, d). We might also lump together those who were optimally matched with
those who were a little too high and a little too low into a 'satisfactory' category (f),
leaving a residual 'unsatisfactory' figure (g) outside these limits (26% unsatisfactory
and 74% satisfactory for 1000 mm). The statistically minded punter should settle for
a working height of 1050 mm.

This is not quite the end of the process since at the best compromise height some
15% of users will have an 'unsatisfactory' match. Is this an acceptable or tolerable
situation or will they be severely uncomfortable or suffer long-term damage? Is it
better to have a bench that is too high or one that is too low? Do we in fact require
an adjustable workbench or some similarly varied solution?

The best possible compromise height is in fact 75 mm below the elbow height of
the average user (i.e. at the mid-point of the optimum range). With the wisdom of
hindsight we can see that this follows necessarily from the shape of the normal
distribution. Having laboriously analysed the problem we find that it could have
been solved by inspection. We could write out our reasoning as follows:

- criterion: elbow height − 100 mm ⩽ bench height ⩽ elbow height − 50 mm;
- best possible compromise: bench height = average elbow height − 75 mm;
- design specification: bench height = 1050 mm.

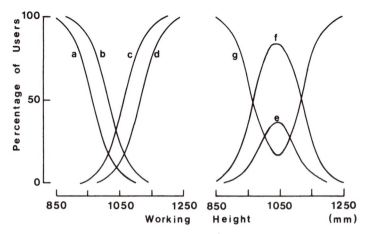

Figure 2.7 The anthropometric method of limits, applied to the determination of the optimal working height, for an industrial assembly task. Curves show the percentage of users accommodated or otherwise: (a) much too low; (b) too low; (c) too high; (d) much too high; (e) just right; (f) satisfactory; (g) unsatisfactory. See text for definition of categories and discussion of technique.

2.4 Anthropometric data

2.4.1 Sources

Few organizations outside the military have the resources to mount a full-scale anthropometric survey. As a consequence, we have extensive and detailed anthropometric data for many of the world's armed services; but relatively few data for the civilian populations from whom they were recruited and of whom they may or may not be representative samples. For example, the only currently available dataset that may be reasonably assumed to be representative of the adult population of the UK, was collected by the Office of Population Censuses and Surveys in 1981 (OPCS 1981; Knight 1984) – and in this survey the only body dimensions measured were height and weight.

Confronted with this situation we may take the purist approach and only quote sources of unimpeachable accuracy; or we may take the pragmatic approach and fill in the gaps as best we can by using various rule-of-thumb methods of estimation (and a certain amount of informed guesswork). I have adopted the latter approach.

There are a number of techniques that you can use to fill the gaps in this way when hard data are not available. These are described in detail in the Appendix. The one I have used most extensively in this book is the method of *ratio scaling*. This technique assumes that although different samples drawn from a particular 'parent' population (as defined in terms of age, sex and ethnic origin) may vary greatly in size, they are likely to be relatively similar in shape. Thus if we have stature data only for the target population that interests us, but we have detailed body-part measurements for an equivalent population or population sample, we may use the former to 'scale up' the latter.

A detailed validation study of this technique is described in Pheasant (1982a). The 1st and 99th %ile values of some 136 dimensions drawn from six different surveys were estimated from a knowledge of only the parameters of stature in the survey concerned. (This is a more rigorous test than the 5th and 95th %iles commonly used in design work.) The errors that accrued were random and conformed approximately to a normal distribution with parameters of −3 [13] mm. Ninety-three per cent of the errors fell within the range of ±25 mm. In many cases the estimates were within the confidence limits of the original survey.

2.4.2 Accuracy

What accuracy is actually required in anthropometric data? This is a very difficult question which must be studied at several different levels. In the purely formalized statistical sense we may consider what percentiles a given percentile that is erroneously quoted actually represent, e.g. if a 95th %ile was in error, the figure quoted might in truth represent the 93rd or 98th %ile, with a consequence of mismatching a greater or lesser percentage of the target population in the design. In the validation study the estimates of the 1st and 99th %iles were checked in this way – on average the estimates would have included 96% of the population as against 98% for perfect data. It is more informative, however, to consider the likely errors of prediction alongside those that might arise in other ways. The human body has very few sharp edges – its contours are rounded and it is generally squashy and unstable. The consequent difficulty in identifying landmarks and controlling posture makes it virtually impossible to achieve an accuracy of better than 5 mm in most anthropometric measures – and for some dimensions the errors may be much worse (sitting elbow height is a notorious example). These errors, however, pale into insignificance in comparison with those that might occur in the application of even the most accurate tables.

In applying anthropometric data we commonly need to make corrections for clothing, postural variation, and so on (see below). These corrections, although they are better than arbitrary, will usually be inexact – as (perhaps more importantly) will be the anthropometric criteria we apply to define a match. Take the case of seat height which we discussed above. The sensations of discomfort which the user experiences will become progressively more pronounced as the height of the seat exceeds his or her popliteal height. But there is no obvious and clearly defined cut-off point at which we should say 'thus far and no further'. In practice, anthropometric criteria are almost always 'fuzzy' in this way.

There are doubtless certain safety-critical applications in which accuracy would be at a premium. But experience indicates that these are the exception rather than the rule. In practice there would be few everyday problems requiring an ergonomic specification to an accuracy of more than 25 mm. We could call this the *anthropometric inch*.

2.4.3 Clothing corrections

Most anthropometric measurements are made on unclothed people; most products and environments are used by clothed people. The data tabulated below are for unclothed people. So before applying these data to any particular problem it will in general be necessary to add an appropriate correction for clothing. (It makes sense to do it this way, rather than to quote figures with clothing corrections already added, because the magnitude of the correction may vary greatly depending upon the circumstances.)

The most important of these corrections is an increment for the heels of shoes which must be added to all vertical dimensions which are measured from the floor. The thinnest pair of carpet slippers has a heel height of only 10 mm. The most outrageous pair of high heels may add 150 mm to a woman's height. A typical heel height for men's ordinary everyday shoes and for women's flats is around 25 mm ± 5.

Women's shoes (and men's shoes to a lesser extent) are subject to periodic changes in fashion. The products and spaces we design will presumably remain in use over several of these fashion cycles. In theory therefore we should base our heel height correction on the midpoint about which these cycles oscillate. In theory also we should add an increment to the standard deviation of our dimension as well as the mean, to allow for variability in heel height. In practice, however, variability in heel height is small compared with anthropometric variability and a uniform increment to all percentiles will be adequate.

Taking one consideration with another, the following corrections would seem appropriate for shoes worn in public places on formal and semi-formal occasions:

- for men, add 25 mm to all dimensions;
- for women, add 45 mm to all dimensions.

Corrections for situations where other types of footwear are the norm should be made on an *ad hoc* basis. Other clothing corrections are in general likely to be small – except for very heavy outdoor clothing or for specialized protective gear, etc. Some examples are given below when discussing individual body dimensions.

2.4.4 Standard anthropometric postures

Most of the measurements described below (and likewise, those given in Chapter 10) were made in one of two standard postures.

In the *standard standing posture* the subject stands erect, pulling himself up to his full height and looking straight ahead, with his shoulders relaxed and his arms hanging loosely by his sides. He stands free of walls, measuring instruments, etc.

In the *standard sitting posture* the subject sits erect on a horizontal, flat surface, pulled up to his full height and looking straight ahead. The shoulders are relaxed, with the upper arms hanging freely by the sides and the forearms horizontal (i.e. the elbows are flexed to a right angle). The height of the seat is adjusted (or blocks are placed under the feet) until the thighs are horizontal and the lower legs are vertical (i.e. the knees are flexed to a right angle). Measurements are made perpendicular to two reference planes. The *horizontal reference plane* is that of the seat surface. The *vertical reference plane* is a real or imaginary plane which touches the back of the uncompressed buttocks and shoulder blades of the subject. The *seat reference point* (SRP) lies at the point of intersection of these two planes and the *median plane* of the body (i.e. the plane that divides it equally into its right and left halves).

People rarely use these upright positions in everyday life. In practice this may not be so much of a problem as it seems, since we shall commonly set our criteria in such a way as to take this into account. There are circumstances, however, where it may be appropriate to make a nominal correction for *normal sitting slump*. Where this is the case, as a rough approximation for adult populations, subtract 40 mm from all percentiles of relevant sitting dimensions.

2.4.5   Defining the target population

The principal factors to take into account when defining a target population of users, for the purpose of selecting an appropriate source of anthropometric data, will in general be: sex, age, nationality (or ethnicity) and occupation (or social class), generally in that order of importance. Where the target population includes children, then age will take first place. The presence of ethnic minorities in a population sample tends to be more of a problem in theory than in practice. As a general rule of thumb, percentile values are unlikely to be affected to any significant extent until the minority group reach 30% of the total or more. Again, however, there may be exceptions for certain safety-critical applications (e.g. guarding of machinery; see Thompson and Booth 1982).

At the end of this chapter you will find a table of best estimate figures for the bodily dimensions of the adult population of the UK aged 19–65 years (Table 2.3). In the chapters that follow we shall treat this as the *standard reference population* on which we shall base our design recommendations and other anthropometric calculations. Data for other target populations and details of sources, etc., will be found in Chapter 10.

2.5   An annotated list of body dimensions

2.5.1   Stature

Definition

The vertical distance from the floor to the vertex (i.e. the crown of the head).

Applications

As a cross-referencing dimension for comparing populations and estimating data; defines the vertical clearance required in the standing workspace; minimal acceptable height of overhead obstructions such as lintels, roofbeams, light fittings, etc.

Corrections

Shoes as above; 25 mm for a hat; 35 mm for a protective helmet. A few design applications call for supine or prone body length (in which the subject lies on his back or front, respectively). Such a position lengthens the adult body by approximately 15 mm.

Notes

(i) In children under 2 years, who cannot stand unaided, crown–heel length is the

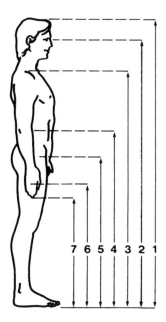

Figure 2.8 Body dimensions.

nearest equivalent dimension. The child must be stretched out in a supine position and prevented from wriggling.

(ii) If you ask an adult person to tell you their height, you must expect them to overestimate by an average of about 25 mm.

2.5.2   Eye height

Definition

Vertical distance from the floor to the inner canthus (corner) of the eye.

Applications

Centre of the visual field; reference datum for location of visual displays as in sections 3.3, 'reach' dimension for sight lines, defining maximal acceptable height of visual obstructions; optical sighting devices for prolonged use should be adjustable for the range of users.

Correction

Shoes as above.

2.5.3   Shoulder height

Definition

Vertical distance from the floor to the acromion (i.e. the bony tip of the shoulder).

Applications

The approximate centre of rotation of the upper limb and, hence, of use in determining zones of comfortable reach; reference datum for location of fixtures, fittings, controls etc.).

Correction

Shoes as above.

2.5.4   Elbow height

Definition

Vertical distance from the floor to the radiale. (The radiale is the bony landmark formed by the upper end of the radius bone which is palpable on the outer surface of the elbow.)

Applications

An important reference datum for the determination of work-surface heights, etc. (see section 3.8).

Correction

Shoes as above.

Note

Some surveys measure to the underside of the elbow when it is flexed to a right angle. This gives a value approximately 15 mm less than the standard measurement.

2.5.5   Hip height

Definition

Vertical distance from the floor to the greater trochanter (a bony prominence at the upper end of the thigh bone, palpable on the lateral surface of the hip).

Applications

Centre of rotation of the hip joint, hence the functional length of the lower limb.

Correction

Shoes as above.

2.5.6 Knuckle height

Definition

Vertical distance from the floor to metacarpal III (i.e. the knuckle of the middle finger).

Applications

Reference level for handgrips; for support (handrails, etc.) approximately 100 mm above knuckle height is desirable. Handgrips on portable objects should be at less than knuckle height. Optimal height for exertion of lifting force (see Section 8.4).

Correction

Shoes as above.

2.5.7 Fingertip height

Definition

Vertical distance from the floor to the dactylion (i.e. the tip of the middle finger).

Application

Lowest acceptable level for finger-operated controls.

Correction

Shoes as above.

2.5.8 Sitting height

Definition

Vertical distance from the sitting surface to the vertex (i.e. the crown of the head).

Applications

Clearance required between seat and overhead obstacles.

Corrections

10 mm for heavy outdoor clothing beneath the buttocks; variable amount for seat compression; 25 mm for a hat; 35 mm for a safety helmet.

2.5.9   Sitting eye height

Definition

Vertical distance from the sitting surface to the inner canthus (corner) of the eye.

Applications

See dimension 2.

Corrections

10 mm for heavy outdoor clothing; up to 40 mm reduction for 'sitting slump'; seat compression.

2.5.10   Sitting shoulder height

Definition

Vertical distance from the seat surface to the acromion (i.e. the bony point of the shoulder).

Applications

Approximate centre of rotation of the upper limb.

Correction

As for dimension 9.

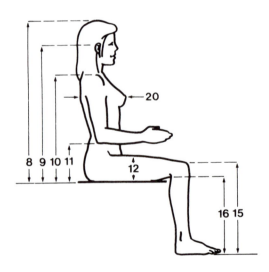

Figure 2.9   Body dimensions.

2.5.11   Sitting elbow height (also known as elbow rest height)

Definition

Vertical distance from the seat surface to the underside of the elbow.

Applications

Height of armrests; important reference datum for the heights of desk tops, keyboards, etc., with respect to the seat (see Section 4.3).

2.5.12   Thigh thickness (also known as thigh clearance)

Definition

Vertical distance from the seat surface to the top of the uncompressed soft tissue of the thigh at its thickest point, generally where it meets the abdomen.

Applications

Clearance required between the seat and the underside of tables or other obstacles (see Section 6.1).

Correction

35 mm for heavy outdoor clothing.

2.5.13   Buttock–knee length

Definition

Horizontal distance from the back of the uncompressed buttock to the front of the kneecap.

Applications

Clearance between seat back and obstacles in front of the knee (see Section 6.1).

Correction

20 mm for heavy outdoor clothing.

2.5.14   Buttock–popliteal length

Definition

Horizontal distance from the back of the uncompressed buttocks to the popliteal angle, at the back of the knee, where the back of the lower legs meet the underside of the thigh.

Applications

Reach dimension, defines maximum acceptable seat depth (see Section 4.3).

2.5.15   Knee height

Definition

Vertical distance from the floor to the upper surface of the knee (usually measured to the quadriceps muscle rather than the kneecap).

Applications

Clearance required beneath the underside of tables, etc. (see Section 6.1).

Corrections

Shoes as above.

2.5.16   Popliteal height

Definition

Vertical distance from the floor to the popliteal angle at the underside of the knee where the tendon of the biceps femoris muscle inserts into the lower leg.

Application

Reach dimension defining the maximum acceptable height of a seat (see Section 4.3).

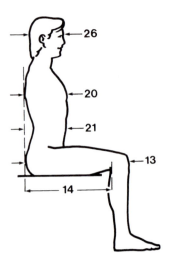

Figure 2.10   Body dimensions.

Correction

Shoes as above.

2.5.17   Shoulder breadth (bideltoid)

Definition

Maximum horizontal breadth across the shoulders, measured to the protrusions of the deltoid muscles.

Applications

Clearance at shoulder level.

Corrections

10 mm for indoor clothing; 40 mm for heavy outdoor clothing.

2.5.18   Shoulder breadth (biacromial)

Definition

Horizontal distance across the shoulders measured between the acromia (bony points).

Applications

Lateral separation of the centres of rotation of the upper limb.

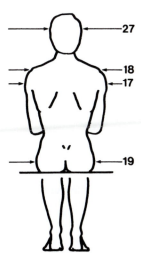

Figure 2.11   Body dimensions.

2.5.19   Hip breadth

Definition

Maximum horizontal distance across the hips in the sitting position.

Applications

Clearance at seat level; the width of a seat should be not much less than this (see Section 4.3).

Corrections

10 mm for light clothing; 25 mm for medium clothing; 50 mm for heavy outdoor clothing.

Note

This is a dimension with a substantial soft tissue component. In studies of physique, etc., the bony dimension bicristal breadth is generally used (measured between the lateral edges of the crests of the hip bones).

2.5.20   Chest (bust) depth

Definition

Maximum horizontal distance from the vertical reference plane to the front of the chest in men or breast in women.

Applications

Clearance between seat backs and obstructions.

Corrections

Up to 40 mm for outdoor clothing.

2.5.21   Abdominal depth

Definition

Maximum horizontal distance from the vertical reference plane to the front of the abdomen in the standard sitting position.

Applications

Clearance between seat back and obstructions.

Corrections

Up to 40 mm for outdoor clothing.

2.5.22 Shoulder–elbow length

Definition

Distance from the acromion to the underside of the elbow in a standard sitting position.

2.5.23 Elbow–fingertip length

Definition

Distance from the back of the elbow to the tip of the middle finger in a standard sitting position.

Applications

Forearm reach; used in defining normal working area (see Section 3.4).

Corrections

For general reach corrections, see dimension 34.

2.5.24 Upper limb length

Definition

Distance from the acromion to the fingertip with the elbow and wrist straight (extended).

2.5.25 Shoulder–grip length

Definition

Distance from the acromion to the centre of an object gripped in the hand, with the elbow and wrist straight.

Applications

Functional length of upper limb; used in defining zone of convenient reach (see Section 3.3).

Corrections

Reach corrections as in dimension 34.

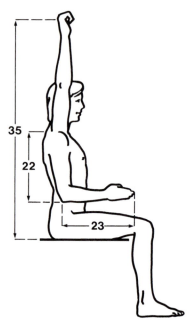

Figure 2.12   Body dimensions.

2.5.26   Head length

Definition

Distance between the glabella (the most anterior point on the forehead between the brow ridges) and the occiput (back of the head) in the midline.

Applications

Reference datum for location of eyes, approximately 20 mm behind glabella (see Section 3.7).

2.5.27   Head breadth

Definition

Maximum breadth of the head above the level of the ears.

Applications

Clearance.

Corrections

Add 35 mm for clearance across the ears; up to 90 mm for protective helmets.

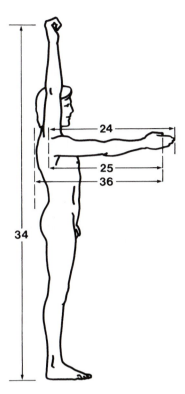

Figure 2.13   Body dimensions.

2.5.28   **Hand length**

Definition

Distance from the crease of the wrist to the tip of the middle finger with the hand held straight and stiff.

Applications

See dimension 34 and Section 5.1.

2.5.29   **Hand breadth**

Definition

Maximum breadth across the palm of the hand (at the distal ends of the metacarpal bones).

Applications

Clearance required for hand access, e.g. grips, handles, etc. (see Sections 5.1 and 5.3).

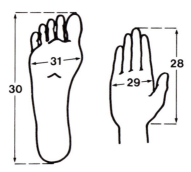

Figure 2.14   Body dimensions.

Corrections

As much as 25 mm for some protective gloves.

2.5.30   Foot length

Definition

Distance, parallel to the long axis of the foot, from the back of the heel to the tip of the longest toe.

Applications

Clearance for foot, design of pedals.

Corrections

In many respects surveys of shoes would be more relevant than surveys of feet since their sizes and shapes are often unrelated. For the purposes of argument, we could add 30 mm for men's street shoes and 40 mm for protective boots.

2.5.31   Foot breadth

Definition

Maximum horizontal breadth, wherever found, across the foot perpendicular to the long axis.

Applications

Clearance for foot, spacing of pedals, etc.

Corrections

See dimension 30; add 10 mm for men's street shoes; 30 mm for heavy boots.

2.5.32   Span

Definition

The maximum horizontal distance between the fingertips when both arms are stretched out sideways.

Application

Lateral reach (see Section 3.2).

Corrections

See dimension 34.

2.5.33   Elbow span

Definition

Distance between the tips of the elbows when both upper limbs are stretched out sideways and the elbows are fully flexed so that the fingertips touch the chest.

Applications

A useful guideline when considering 'elbow room' in the workspace.

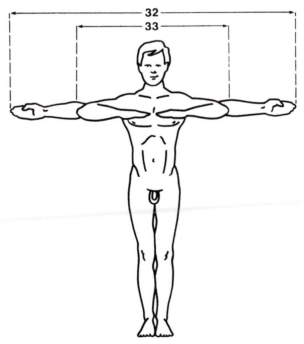

Figure 2.15   Body dimensions.

2.5.34–36   Grip reaches

Definitions

In each case the measurement is made to the centre of a cylindrical rod fully grasped in the palm of the hand. In dimensions 34 and 35 the arm is raised vertically above the head and the measurement is made from the floor or seat surface, respectively. In dimension 36 the arm is raised horizontally forward at shoulder level and the measurement is taken from the back of the shoulder blades. In each case these are 'easy' reaches made without excessive stretch.

Corrections

Some surveys measure reach to the tip of the outstretched middle finger or to the tip of the thumb when it forms a 'pinch' with the index finger. Approximately,

- fingertip reach = grip reach + 60% hand length;
- thumbtip reach = grip reach + 20% hand length.

See Section 3.2 for a discussion of reach envelopes in general.

Table 2.3 Anthropometric estimates for British adults aged 19–65 years (all dimensions in mm, except for body weight, given in kg).

Dimension	Men				Women			
	5th %ile	50th %ile	95th %ile	SD	5th %ile	50th %ile	95th %ile	SD
1. Stature	1625	1740	1855	70	1505	1610	1710	62
2. Eye height	1515	1630	1745	69	1405	1505	1610	61
3. Shoulder height	1315	1425	1535	66	1215	1310	1405	58
4. Elbow height	1005	1090	1180	52	930	1005	1085	46
5. Hip height	840	920	1000	50	740	810	885	43
6. Knuckle height	690	755	825	41	660	720	780	36
7. Fingertip height	590	655	720	38	560	625	685	38
8. Sitting height	850	910	965	36	795	850	910	35
9. Sitting eye height	735	790	845	35	685	740	795	33
10. Sitting shoulder height	540	595	645	32	505	555	610	31
11. Sitting elbow height	195	245	295	31	185	235	280	29
12. Thigh thickness	135	160	185	15	125	155	180	17
13. Buttock–knee length	540	595	645	31	520	570	620	30
14. Buttock–popliteal length	440	495	550	32	435	480	530	30
15. Knee height	490	545	595	32	455	500	540	27
16. Popliteal height	395	440	490	29	355	400	445	27
17. Shoulder breadth (bideltoid)	420	465	510	28	355	395	435	24
18. Shoulder breadth (biacromial)	365	400	430	20	325	355	385	18
19. Hip breadth	310	360	405	29	310	370	435	38
20. Chest (bust) depth	215	250	285	22	210	250	295	27
21. Abdominal depth	220	270	325	32	205	255	305	30
22. Shoulder–elbow length	330	365	395	20	300	330	360	17
23. Elbow–fingertip length	440	475	510	21	400	430	460	19
24. Upper limb length	720	780	840	36	655	705	760	32
25. Shoulder–grip length	610	665	715	32	555	600	650	29
26. Head length	180	195	205	8	165	180	190	7
27. Head breadth	145	155	165	6	135	145	150	6
28. Hand length	175	190	205	10	160	175	190	9
29. Hand breadth	80	85	95	5	70	75	85	4
30. Foot length	240	265	285	14	215	235	255	12
31. Foot breadth	85	95	110	6	80	90	100	6
32. Span	1655	1790	1925	83	1490	1605	1725	71
33. Elbow span	865	945	1020	47	780	850	920	43
34. Vertical grip reach (standing)	1925	2060	2190	80	1790	1905	2020	71
35. Vertical grip reach (sitting)	1145	1245	1340	60	1060	1150	1235	53
36. Forward grip reach	720	780	835	34	650	705	755	31
Body weight	*55*	*75*	*94*	*12*	*44*	*63*	*81*	*11*

See notes on pp. 30–44.

CHAPTER THREE

Workspace design

In this chapter we shall consider the design and layout of the spaces in which people live and work, with particular reference to anthropometric considerations of:

- reach;
- clearance;
- posture.

All dimensional data given in this chapter are for the standard reference population as described in Table 2.3.

The rationalization of workspace layout is partly a matter of anthropometrics and partly a matter of common sense. The commonsense element is embodied in the four principles set out in Table 3.1 – which were first stated in a formal way by the late Ernest J. McCormick (1970). These principles are applicable to a large class of design problems which involve considerations of 'what to put where': the controls and displays on a panel, the furniture and appliances in a kitchen or the machines on a shop floor, the facilities in a large building and so on – and even perhaps to more abstract problems like the arrangement of information in a database.

Table 3.1 Principles of rational workspace layout.

- Importance principle – the most important items should be in the most accessible locations
- Frequency-of-use principle – the most frequently used items should be in the most accessible locations
- Function principle – items with similar functions should be grouped together
- Sequence-of-use principle – items that are commonly used in sequence should be laid out in the same sequence

Source: after McCormick (1970)

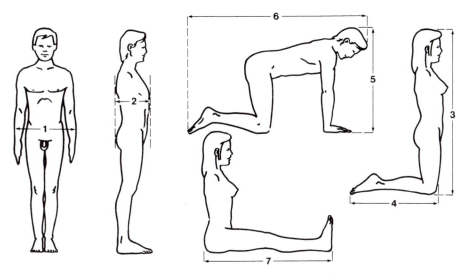

Figure 3.1   Clearance dimensions in various positions as given in Table 3.2.

3.1   Clearance

Table 3.2 and Figure 3.1 present clearance data for a variety of working positions, derived from a variety of sources (Damon *et al.* 1966, Van Cott and Kinkade 1972, Department of Defense 1981) and scaled, as far as possible, to match the standard population. The maximum breadth and depth of the body are overall measurements taken at the widest or deepest point wherever this occurs. The male data, based on US servicemen, exceed any relevant dimensions in Table 2.3 and have been quoted direct.

Fruin (1971), in the context of an account of pedestrian movement and flow, introduced the concept of the body ellipse. In plan view the space occupied by the human body may be approximately described by an ellipse – the long and short axes of which are determined by its maximum breadth and depth. Taking the 95th %ile male data from Table 3.2 and allowing a generous 25 mm all round for clothes, the long and short axes of our ellipse become 630 and 380 mm, respectively. Figure

Table 3.2   Clearance dimensions in various positions (all dimensions in mm).

Dimension	Men				Women			
	5th %ile	50th %ile	95th %ile	SD	5th %ile	50th %ile	95th %ile	SD
1. Maximum body breadth	480	530	580	30	355	420	485	40
2. Maximum body depth	255	290	330	22	225	275	325	30
3. Kneeling height	1210	1295	1380	51	1130	1205	1285	45
4. Kneeling leg length	620	685	750	40	575	630	685	32
5. Crawling height	655	715	775	37	605	660	715	33
6. Crawling length	1215	1340	1465	75	1130	1240	1350	66
7. Buttock–heel length	985	1070	1160	53	875	965	1055	55

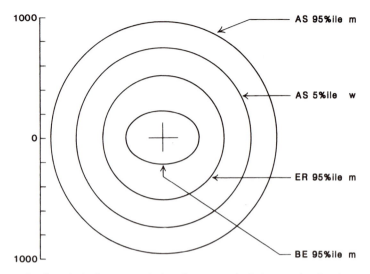

Figure 3.2 Body ellipse (BE), elbow room (ER) and arm span (AS). See text for details.

3.2 shows this ellipse. To give us some idea of 'elbow room', a circle has been drawn around the ellipse. The diameter of this circle is the 95th %ile male elbow span (1020 mm). Two further circles, the diameters of which are determined by the arm span of a 5th %ile woman and a 95th %ile man, complete a first simple analysis of space requirements.

3.1.1 Whole body access

Table 3.3 and Figure 3.3 present figures for minimum dimensions for hatches, etc., giving whole-body access to and egress from confined spaces – as given in various sources. MIL-STD-1472C (a US military standard) also specifies that the maximum 'step down' distance from a horizontal hatch should be 690 mm.

ISO 2860 (an International Standard dealing with earth-moving machinery) states that the dimensions it gives are 'the smallest that will accommodate 95% of the worldwide operator population'. And comparison of the MIL-STD-1472C figures with the data of Table 3.1 suggests that these are also based on the 95th %ile values of the dimensions concerned.

By inference this means that 5% of men trying to pass through hatches of these dimensions would get stuck – which in a safety-critical application would clearly be unacceptable. Access to things like pressure vessels present a particular problem in this respect, since the aperture must be large enough to permit emergency evacuation (perhaps by two people carrying a stretcher, etc.), but as small as possible so as not to compromise structural strength. I recently encountered a similar problem in the design of whole-body scanners for hospitals.

If we assume that the data for body breadth and body depth in Table 3.2 are normally distributed – and we apply the equation for calculating percentiles given in section 2.1 – we may calculate that:

— a 580 × 330 aperture excludes approximately 1 man in 20

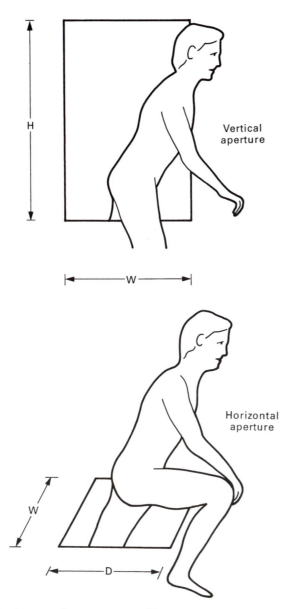

Figure 3.3 Whole-body access dimensions; see Table 3.3.

— a 600 × 340 aperture excludes approximately 1 man in 100
— a 625 × 360 aperture excludes approximately 1 man in 1000
— a 640 × 370 aperture excludes approximately 1 man in 10 000
— a 660 × 385 aperture excludes approximately 1 man in 100 000

Note that these figures do not allow for clothing, personal equipment, etc. The numbers given for the percentage of men excluded may well be underestimates, since the distributions for the dimensions concerned are likely to be positively skewed.

Table 3.3   Minimum dimensions for whole-body access (all dimensions in mm).

For access through an aperture in a horizontal surface (i.e. floor or ceiling)

Source	Rectangular aperture width (W) × depth (D)	Circular aperture Diameter
Light clothing		
ISO 2860	560 × 330	560
MIL-STD-1472C	580 × 330	
Bulky clothing		
ISO 2860	650 × 470	600
MIL-STD-1472C	690 × 410	760
Damon *et al.* (1966)	740 × 510	

For access through an aperture in a vertical surface (i.e. wall)

Source	Rectangular aperture height (H) × width (W)
Light clothing	
MIL-STD-1472C	660 × 760
Heavy clothing	
MIL-STD-1472C	740 × 860
Damon *et al.* 1966	780 × 500

Note: Data from Damon *et al.* (1966) are also quoted in Van Cott and Kincade (1972) and Woodson (1981).

Table 3.4   Minimum dimensions for passageways in areas of restricted access.

	Height (mm)	Width (mm)
Walking[a]		
upright	1955[b]	
stooped	1600	
straight ahead		630
crabwise (sideways)		380
Crawling on hands and knees	815	630
Crawling prone[c]	430	630

Notes:
[a] for walking a trapezoidal space which is 630 mm wide at shoulder height and 145 mm wide at floor level will suffice.
[b] stature of a 99th %ile man wearing shoes and protective helmet.
[c] for prone crawling a width of 1015 mm is preferable to allow for lateral elbow movements.
Source: data based on Damon *et al.* (1966), with modifications.

Table 3.5 Space requirements for circulation (all dimensions in mm).

Widths of access		
One person walking normally	650 (600 restricted)	
Two people passing or walking side by side	1350 (1200 restricted)	
One person walking, another flattened against wall	1000 (900 restricted)	
Two people passing crabwise	900 (850 restricted)	
One person carrying a suitcase	800	
One person carrying a tea-tray	900	
One person carrying two suitcases	1000	
One person with a raised umbrella	1150	
Two people with raised umbrellas	2350	
One person with crutches	840	
One person with a walking frame	1000	
Wheelchair user – minimum	750	
Wheelchair user – reasonable	800	
Wheelchair user – preferred	900	
Passage between obstacles	*Normal*	*Crabwise*
Both obstacles greater than 1000 mm in height	600	400
One obstacle greater than 1000 mm in height, the other less	600	400
Both obstacles less than 1000 mm in height	550	350
Standing in line	450 per person	

Source: after Pheasant (1987).

In the case of emergency exits and escape hatches we should expect speed of passing through to be a function of aperture size up to some critical dimension at which no further improvement was possible. Roebuck and Levendahl (1961) studied the emergency exits of aircraft and found that speed levelled off at a door width of around 510 mm (unless steps were also involved, in which case greater widths were optimal).

Minimum dimensions for passageways in situations of limited access, such as tunnels and catwalks, etc., are given in Table 3.4 (quoted from Damon *et al.* 1966 with slight modifications, introduced for conformity with figures given elsewhere).

3.1.2 Circulation space

Minimum dimensional requirements for circulation space in buildings, passage between obstacles, and so on are summarized in Table 3.5 – which is quoted from Pheasant (1987) – which in turn was based on Tutt and Adler (1979) and Noble (1982).

3.2 Reach – the workspace envelope

Consider what happens when you reach your arm forwards. First, you raise your upper limb through 90° – this is achieved principally by a rotational movement of the arm in its socket on the shoulder blade or scapula (i.e. flexion of the gleno-humeral or true shoulder joint; see Figure 3.10). It is, however, impossible to make such a movement without a small shift in the scapula's position on the chest wall. (Anatomists call this interaction 'scapulo-humeral rhythm'.) You have now reached

Table 3.6   Increments to forward grip reach (all dimensions in mm).

	Men			Women		
Dimension	5th %ile	50th %ile	95th %ile	5th %ile	50th %ile	95th %ile
Basic dimension						
Forward grip reach[a]	720	780	835	650	705	755
Increments						
for a pinch grip (to the thumbtip)	35	40	40	30	35	40
for fingertip operation	105	115	125	95	105	115
for a forward thrust of the shoulder[b]	115	130	150	95	115	140
for 10° of trunk inclination	80	85	95	75	85	95
for 20° of trunk inclination	155	170	185	150	170	185
for 30° of trunk inclination	230	250	270	225	245	270

Notes:
[a] quoted directly from Table 2.3.
[b] calculated from data in MIL-STD-1472c.

the position in which the 'static' dimensions of forward reach, dimension 36 in Table 2.3, would be measured – if your shoulder blades had been touching a wall at the outset they would still be doing so. As you reach farther forwards from this basic starting point, several new movements occur. Your whole shoulder girdle is thrust forwards (protracted) and you begin to incline your trunk forwards by flexion of the hip joint and spine. What determines the final limit of your forward reach? Try it and you will quickly discover that it is the tendency to topple over as the horizontal position of your centre of gravity reaches the limit of the base of support of your feet. This in turn can be modified by pushing the pelvis backwards as a counterbalance, etc. Suitable increments to the basic dimension of forward grip reach are given in Table 3.6.

Dynamic reach may best be characterized by the three-dimensional coordinates of a volume of space. Such a volume is referred to as a 'workspace envelope' (or more grandly as a 'kinetosphere'). Since standing reach is essentially a matter of body equilibrium, the envelope will be modified by any factor that affects this. A weight in the hands will diminish reach. Grieve and Pheasant (1982) reported experiments showing how reach was increased by increasing the footbase and diminished by placing an obstacle behind the subject to limit the activities of counterbalancing.

Several studies of the workspace envelope of the sitting person have been published. That of Kennedy (1964) has been particularly widely quoted (Damon *et al.* 1966, Van Cott and Kinkade 1972, NASA 1978). The reader should note that all reach envelopes are highly specific to the situation in which they were measured. The data of Kennedy (1964) were measured in an aircraft seat with the subjects securely strapped in; had the seat or the restraints been otherwise, the results would have been numerically different.

3.3   Zones of convenient reach

At this point it is appropriate to develop the concept of a zone or space in which an object may be reached conveniently, that is, without undue exertion. Consider what

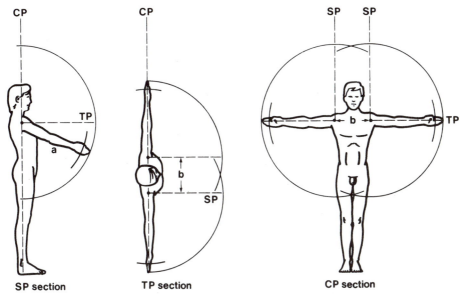

SP section   TP section   CP section

Figure 3.4 Zones of convenient reach (ZCR) seen in elevation and plan. Left to right: vertical section in sagittal plane (SP) passing through shoulder joint; horizontal section in transverse plane (TP) passing through shoulder joints; vertical section in coronal plane (CP) passing through shoulder joints. Each plane of section is marked on the other two diagrams.

it means for a control to be 'within arm's length'. The upper limb, measured from the shoulder to the fingertip (or to the centre of grip), sweeps out a series of arcs centred upon the joint (see Figure 3.4). These define the zone of convenient reach (ZCR) for one hand, which extends sideways to the coronal plane of the body. The

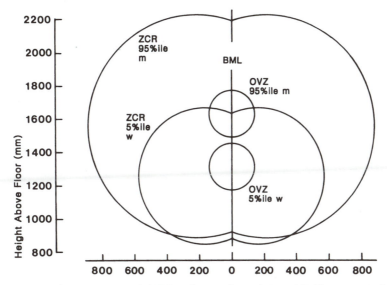

Figure 3.5 Zones of convenient reach (ZCR) and optimal visual zones (OVZ) on a vertical surface 500 mm in front of the shoulders. Ninety-fifth %ile man (m) and 5th %ile woman (w). (BML = body mid-line.)

Table 3.7 Zones of convenient reach[a] (all dimensions in mm).

	Radius (r)					
	Men			Women		
	5th	50th	95th	5th	50th	95th
d	%ile	%ile	%ile	%ile	%ile	%ile
0	610	665	715	555	600	650
100	600	655	710	545	590	645
200	575	635	685	520	565	620
300	530	595	650	465	520	575
400	460	530	595	385	445	510
500	350	440	510	240	380	415
600	110	285	390			250
	Men[b]			Women[b]		
	5th	50th	95th	5th	50th	95th
	%ile	%ile	%ile	%ile	%ile	%ile
Biacromial breadth	365	400	430	325	355	385
Shoulder height (standing, shod)	1340	1425	1560	1260	1335	1450
Shoulder height (sitting)	540	595	645	505	555	605

Notes:
[a] To construct a zone of convenient reach in a vertical plane, a distance d in front of the shoulders, draw two circles of radius r; the centres of the circles are defined by standing or sitting shoulder height and biacromial breadth. To construct a zone of convenient reach in a horizontal plane, distance d above or below the shoulders, draw two semicircles of radius r, centred upon the position of the shoulders.
[b] Figures calculated from equation 3.1 assuming a full grip.

zones for the two limbs intersect in the midline (median) plane of the body. The volume which is thus defined comprises two intersecting hemispheres. The radius of each hemisphere is the upper limb length (a) and their centres are a distance (b) equal to biacromial breadth apart.

Many design problems are concerned with the intersection of vertical, horizontal or (very occasionally) oblique planes with either the volume of the workspace envelope or that of the zone of convenient reach. Suppose we wish to locate a set of items upon the vertical wall of a control room so that they might be conveniently operated by a standing person. The intersection of a plane with a sphere produces a circle. The radius of this circle may be calculated by Pythagoras's theorem as

$$r = \sqrt{a_2 - d_2} \qquad\qquad (3.1)$$

where r is the radius of the circle on the wall, a is the upper limb length (or shoulder grip length) and d is the horizontal distance between the shoulder and the wall. Figure 3.5 shows the construction of such a zone for fingertip controls by a 95th %ile male or a 5th %ile female operator, assuming $d = 500$. The design also involves visual questions – optimal zones for visual displays have been added according to the criteria of section 3.7 below.

The zone of convenient reach may be similarly described for any other vertical or horizontal plane parallel to a line joining the shoulders. Requisite data are given in Table 3.7.

3.4 **The normal working area**

The intersection of a horizontal plane, such as a table or bench, with the zone of convenient reach defines what workstudy writers would call the maximum working area (Barnes 1958). Within this is a much smaller 'normal working area' – described by a comfortable sweeping movement of the upper limb, about the shoulder with the elbow flexed to 90° or a little less. Das and Grady (1983) have discussed this latter at length. The presentation that ensues (Figures 3.6 and 3.7), is based on the original concept of Squires (1956).

A person sits at a bench or table. His shoulder joints are located at S_1 and S_2 which are a distance b apart = biacromial breadth. The elbows are located at E_1 and E_2, at the table's edge, a distance d in front of the shoulders, such that

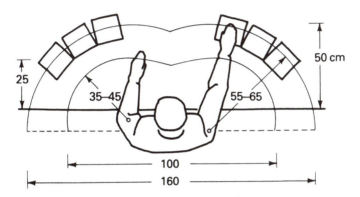

Figure 3.6   Horizontal arc of grasp, and working area at tabletop height. The grasping distance takes account of the distance from shoulder to hand; the working distance only elbow to hand. The values include the 5th %ile, and so apply to men and women of less than average height. (From E. Grandjean, *Fitting the Task to the Man*, Taylor & Francis, 1988, 4th Edn, fig. 42, p. 51; reproduced with kind permission.)

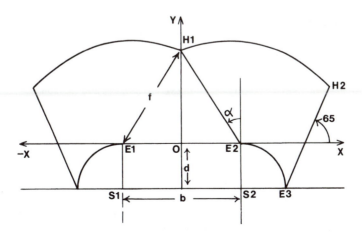

Figure 3.7   Construction of the normal work area (NWA).

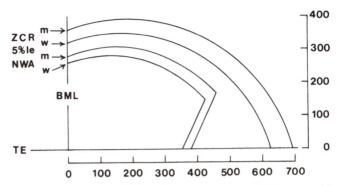

Figure 3.8   Zones of convenient reach (ZCR) and normal work area (NWA) on a table surface, for a 5th %ile man (m) and woman (w) (BML = body mid-line; TE = table edge).

d = abdominal depth/2. Both hands commence at H_1, the location of which is calculated from the dimension f = elbow–grip length. Hence the angle a is given

$$\sin a = b/2f \tag{3.2}$$

When the elbow is flexed at 90° and the humerus is rotated at the shoulder about its own axis, the comfortable limit of outward rotation is limited to about 25°. In the present case the outer limit of the normal working area is a prolate epicycloid, H_1H_2, formed by two simultaneous rotations. The forearm (f) rotates through $a + 25°$, whilst the elbow itself moves outwards and backwards through a circular 90° arc from E_1 to E_3. Hence, the forearm comes to lie at an angle of $90 - 25 = 65°$

Table 3.8   Co-ordinates of the normal working area.[a]

Position	Degrees	5th %ile man		5th %ile woman	
		X	Y	X	Y
H_1	0	0	281	0	257
	10	56	298	53	271
	20	114	307	105	278
	30	172	307	160	279
	40	227	300	211	272
	50	281	287	260	258
	60	333	266	307	237
	70	370	239	350	211
	80	423	206	388	181
H_2	90	460	169	421	146
I	–	380	0	354	0

Note:
[a] Origin is at the table's edge in the midline of the body. The X axis runs along the table's edge; the Y axis is perpendicular. I is the point where the normal working area intersects the table's edge.

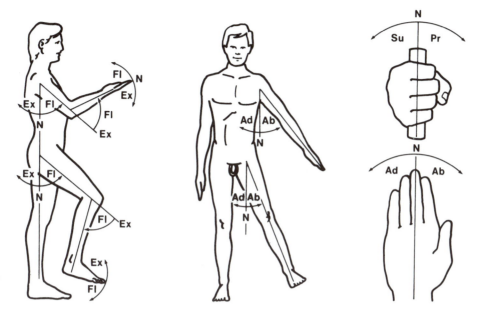

Figure 3.9 Terms used in the description of movements and joint ranges as given in Table 3.9.
Fl = flexion; Ex = extension; Ab = abduction; Ad = adduction; Su = supination; Pr = pronation;
N = neutral position.

to the table edge. Thus, when the arc SE has rotated through $\gamma°$, the arc EH has rotated through $\beta°$ such that

$$\gamma = \beta(a + 25)/90 \qquad (3.4)$$

The co-ordinates of the elbow with respect to the shoulder are given by

$$X_1 = d \sin \beta \qquad (3.5)$$

$$Y_1 = d \cos \beta \qquad (3.6)$$

The co-ordinates of the hand with respect to the elbow are given by

$$X_2 = -f \sin(a - \gamma) \qquad (3.7)$$

$$Y_2 = f \cos(a - \gamma) \qquad (3.8)$$

Therefore the co-ordinates of the hand with respect to a point on the table's edge in the midline of the body are given by

$$X = d \sin \beta - f \sin(a - \gamma) + b/2 \qquad (3.9)$$

$$Y = d \cos \beta + f \cos(a - \gamma) - d \qquad (3.10)$$

Figure 3.9 and Table 3.8 are based on these equations together with the anthropometric data of Table 2.3. (The figure for d is based on a 50th %ile, because abdominal depth is poorly correlated with limb lengths.)

3.5 Joint ranges

The flexibility of the human body is measured in terms of the angular range of motion of the joints. Joint movements are the subject of a terminology which is almost standardized (see Figure 3.9). Consider a vertical plane cutting the body down the midline into equal right and left halves – this is called the median (sagittal) plane. Any vertical plane parallel to it is called a sagittal plane and any vertical plane perpendicular to it is called a coronal plane. In general, sagittal plane movements of the trunk or limbs are called flexion and extension. (Flexion movements are those that fold the body into the curled-up foetal position.) Coronal plane movements are called abduction and adduction. (Abduction movements take a limb segment away from the midline.) Limb segments may also rotate about their own axes – inward (medially) or outward (laterally). Inward rotation of the forearm (turning the palm downwards) is called pronation; outward rotation (turning the palm upwards) is called supination.

There are surprisingly little joint range data available. Table 3.9 is based on a survey of male US servicemen conducted by Dempster (1955), re-analysed by Barter et al. (1957) and quoted extensively (Damon et al. 1966, and elsewhere). Note that measurements were not necessarily made in the postures shown in Figure 3.9 (refer to Damon et al. 1966 for details).

In general, women have a somewhat greater flexibility than men (by about 5–15% on average). Decrements with age are probably small in the absence of joint

Table 3.9   Joint ranges (degrees).

Joint	5th %ile	50th %ile	95th %ile	SD
1. Shoulder flexion	168	188	208	12
2. Shoulder extension	38	61	84	14
3. Shoulder abduction[a]	106	134	162	17
4. Shoulder adduction	33	48	63	9
5. Shoulder medial rotation	61	97	133	22
6. Shoulder lateral rotation	13	34	55	13
7. Elbow flexion	126	142	159	10
8. Pronation[b]	37	77	117	24
9. Supination[c]	77	113	149	22
10. Wrist flexion	70	90	110	12
11. Wrist extension	78	99	120	13
12. Wrist abduction (radial deviation)	12	27	42	9
13. Wrist adduction (ulnar deviation)	35	47	59	7
14. Hip flexion[d]	92	113	134	13
15. Hip abduction	33	53	73	12
16. Hip adduction	11	31	51	12
17. Knee flexion	109	125	142	10
18. Ankle flexion (plantar flexion)	18	38	58	12
19. Ankle extension (dorsiflexion)	23	35	47	7

Notes:
[a] accessory movements of spine increase this to 180°.
[b] rotation of the forearm about its own axis such that the palm faces downwards.
[c] rotation of the forearm about its own axis such that the palm faces upwards.
[d] measured with the knee fully flexed. If the knee is extended the range will be very much less (approx. 60°).
Source: Data from Barter et al. (1957).

disease (osteoarthritis, etc.) but since this is extremely common (universal in some populations) it is reasonable to assume greatly reduced flexibility in the elderly. Unfortunately, few statistical data are available.

The flexibility of one joint may be influenced by the posture of adjacent joints – the most important example is flexion of the hip which is very much greater when the knee is flexed than when it is extended. (Prove this by touching your toes.)

3.6   Working posture

The posture that a person adopts when performing a particular task is determined by the relationship between the dimensions of the person's body and the dimensions of the various items in his or her workspace (a tall person using a standard kitchen will stoop more than a short one, etc.). The extent to which posture is constrained in this way is dependent upon the number and nature of the connections between the person and the workspace. These connections may be either physical (seat, worktop, etc.) or visual (location of displays, etc.). If the dimensional match is inappropriate the short- and long-term consequences for the well-being of the person may be severe.

Posture may be defined as the relative orientation of the parts of the body in space. To maintain such an orientation over a period of time, muscles must be used to counteract any external forces acting upon the body (or in some minority of cases internal tensions within the body). The most ubiquitous of these external forces is gravity. Consider a standing person who leans forwards from the waist. The postural loadings on the hip extensor or the back extensor muscles are proportional to the horizontal distance between the hip and lumbosacral joints, respectively, and the centre of gravity of the upper part of the body (i.e. the head, arms and trunk). The further the trunk is inclined the greater this distance becomes (Figure 3.10). Physiologists call the muscular activity that results from this loading 'static work'.

Muscle as a tissue responds badly to prolonged static mechanical loading. (The same is probably true of other soft tissues, and even perhaps of bone, but the physiology of these cases is much less well understood.) Static effort restricts the flow of blood to the muscle. The chemical balance within the muscle is disturbed, metabolic waste products accumulate and the condition of 'muscular fatigue' supervenes. The person experiences a discomfort which is at first vague but which subsequently develops into a nagging pain until it becomes a matter of some urgency that relief is sought by a change of position. Should you require evidence of this course of events, you should raise one of your arms and hold it out in front of you as you continue to read (or attempt to do so). Provided our workspace and/or working schedule allows us to make the frequent shifts of posture which are subjectively desirable, all will be well – since the physiological processes of muscular fatigue are relatively rapidly reversible by rest or change of activity (particularly if the activity involves stretching the fatigued muscle).

In general, we may think of 'fidgeting' as our bodies' defence against postural stress. This mechanism characteristically operates at a subconscious level – usually we fidget before we become consciously aware of discomfort. In relaxed sitting the sensory stimuli probably come more from the compression of the soft tissues of the buttocks and thighs than from muscle tension. The crossing and uncrossing of the

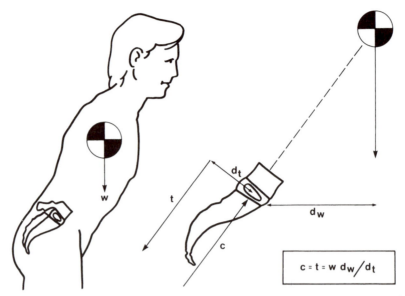

$$c = t = w\, d_w / d_t$$

Figure 3.10 Biomechanical analysis of postural stress in a forward leaning position. Note that this analysis ignores the direct effect of the weight of the trunk which the spine must support even when $d_w = 0$. w is weight of that part of the body above the lumbo-sacral joint; c is the compressive force acting along the axis of the spine; t is the tension in the back muscles (erector spinae).

legs is a characteristic way of redistributing the pressure on the buttocks and, hence, pumping blood through the tissues. The rate of fidgeting can be used as an index of the comfort of chairs – the less comfortable we are, the more we fidget. It is a matter of common experience that other factors are involved. Some people fidget more than others and we all fidget more when we are bored – presumably because mental activity can 'shut out' the sensory stimuli that cause the fidgeting (or raise our threshold of discomfort). Such a hypothesis is in line with contemporary theories of the nature of pain (Melzack and Wall 1982). Students almost universally consider lecture theatre seating to be uncomfortable – is this to do with the seats or the lectures?

Physiologically, comfort is the absence of discomfort – I know of no nerve endings capable of transmitting a positive sensation of comfort from a chair. Comfort is a state of mind which results from the absence of unpleasant bodily sensations. (The same relationship does not hold, however, for pleasure and pain.) We shall consider the matter of sitting comfort at greater length in the next chapter.

Suppose that the working circumstances are such as to closely constrain us to a particular posture and prevent postural change – the consequences may be divided into those occurring over the short term and those occurring over the long term. In the short term, mounting discomfort may distract the operator from his task leading to an increased error rate, reduced output, accidents, etc. From the physiological standpoint, however, we are still talking about a reversible state – since the symptoms are relieved by rest or by a change of activity. At some point, nevertheless (and this point is not well defined since the transition is probably gradual rather than sharp), pathological changes in the muscle or soft tissue take over. Typically, pain

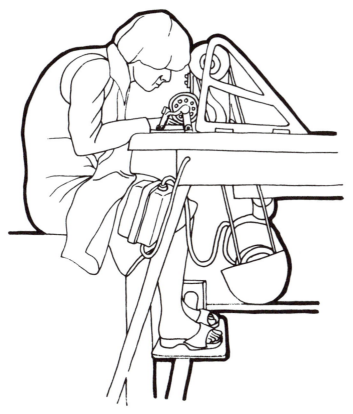

Figure 3.11 The sewing machinist, from an original kindly supplied by Murray Sinclair. (From S. Pheasant, *Ergonomics, Work and Health*, Macmillan, 1991, fig. 1.6, p. 12, reproduced with kind permission.)

comes on after increasingly short periods of postural loading and rest is less certain to bring relief. At this point we are dealing not with discomfort but with physical injury and a disease process.

Back pain, neck pain and the class of conditions affecting the hand, wrist and arm which we refer to as *work-related upper limb disorders* (WRULD) or *repetitive strain injuries* (RSI) are all conditions that characteristically result from over-use of the muscles and other soft tissues in question. This over-use may be due to prolonged static loading, repetitive motions, acute over-exertion or some combination of these. Psychological factors may also be involved (probably because psychological stress leads to increased muscle tension). We shall return to these matters in Chapter 8.

In general, a varied working posture is better than a fixed working posture; but if circumstances demand that you work in a fixed position (as in practice will very often be the case), then the deleterious effects that ensue will increase with the degree of static work required to maintain the position concerned. The following simple guidelines are based in part upon Corlett (1983); for a more detailed discussion see Pheasant (1991a).

(i)   Encourage frequent changes of posture

Sedentary workers, therefore, should be able to sit in a variety of positions – some office chairs are now being designed with this in mind. For many industrial tasks a sit–stand workstation is to be advocated. The task is typically set at a height that is suitable for a standing person (see Section 3.8) and a high stool or 'perch' is provided as an alternative. There seems little doubt that most sedentary workers would be better off if their jobs required them to get up and move around once in a while.

(ii)   Avoid forward inclination of the head and trunk (Figure 3.11)

This commonly results from visual tasks, machine controls or working surfaces that are too low (see below).

(iii)   Avoid causing the upper limbs to be held in a raised position (Figure 3.12)

This commonly results from a working level that is too high (or a seat that is too low). If manipulative tasks must be performed in a raised position, perhaps for visual reasons, arm supports should be provided. In addition to being a considerable stress to the shoulder muscles, tasks that must be performed at above the level of the heart impose an additional circulatory burden. The upper limit for manipulative tasks should be around halfway between elbow and shoulder level.

Figure 3.12   Deviated wrist positions in repetitive industrial tasks, showing movements of radial and ulnar deviation with an extended wrist in a packing task, where the working level is too high. Note also the abduction of the shoulders. From an original kindly supplied by Peter Buckle. (From S. Pheasant, *Ergonomics, Work and Health*, Macmillan, 1991, fig. 14.1, p. 262, reproduced with kind permission.)

(iv) **Avoid twisted and asymmetrical positions**

These commonly result from expecting an operator to have eyes in the back of his head, i.e. from the mislocation of displays and controls.

(v) **Avoid postures that require a joint to be used for long periods at the limit of its range of motion**

This is particularly important for the forearm and wrist.

(vi) **Provide adequate back support in all seats**

It may be that for operational reasons the backrest cannot be used during the performance of the work task – but it will still be important in the rest pauses.

(vii) **Where muscular force must be exerted the limbs should be in a position of greatest strength**

Unless by so doing one of the foregoing rules is broken (see Section 3.9).

3.7 **Vision and the posture of the head and neck**

The visual demands of a task and the location of visual displays are important not only in themselves, but also because they largely determine the posture of the head and neck. Look carefully at the printed text on this page – fix your eyes on one particular word near the centre of the page. You will find that other words become less distinct with increasing distance from the central point of fixation and the margins of the page are no more than an indistinct blur. Only the central part of the visual field is sufficiently sensitive for demanding visual tasks such as reading text or recognizing a face. The area of foveal vision, as this central region is called, is limited to a solid angle of some 5° about the line of central fixation. Visual work demands that the foveal regions of both eyes be directed convergently upon the task. Furthermore, the lenses of the eyes must accommodate (focus) to the appropriate distance. The processes of direction and convergence of gaze are integrated with accommodation by a set of flexes so finely tuned that we are unaware of their existence until such times as they break down by reason of age or inebriation.

If we sit or stand with our head up, and look ahead, our eyes will naturally assume a slight downward gaze of some 10 or 15° from the vertical – this we shall call the relaxed line of sight. The direction of gaze is altered, first, by movements of the eyeballs in their sockets (orbits) by means of the orbital muscles, and, second, by movements of the head and neck. Taylor (1973) states that the eyes may be raised by 48° and lowered by 66° without head movements. In practice, only a part of this range of movement is used. Weston (1953), in his classic study of visual fatigue, suggests that, in practice, downward eye movements were limited to 24–27°; beyond

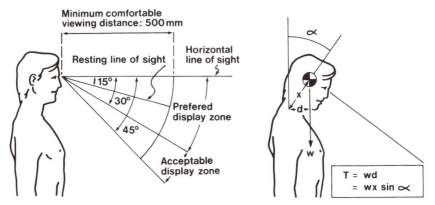

Figure 3.13   Left: preferred viewing conditions as described in text. Right: postural stress to neck muscles resulting from a downward line of sight. *T* is the torque about the neck; *w* is the weight of the head and neck; *x* is the distance from C7 to the centre of gravity of the head and neck.

that point the head and neck are inclined forwards and the neck muscles come under tension to support the weight of the head (see Figure 3.13).

Grandjean *et al.* (1984) described an experiment in which a group of VDU operators were given an adjustable workstation and encouraged to set it to their own satisfaction over a period of 1 week. The preferred visual angle was 9° [4.5°] downwards from the horizontal. Brown and Schaum (1980) have also conducted fitting trials on VDU workstations. Their results are reported in co-ordinate form, but it is possible to calculate that the average preferred visual angle was 18° downwards.

On the basis of the above findings we may conclude that the preferred zone for the location of visual displays extends from the horizontal line of sight downwards to an angle of 30° and that the optimal line of sight is somewhere in the middle of this zone. Given that some modest degree of neck flexion is acceptable, this could be extended a further 15° (see Figure 3.13).

Visual comfort and satisfactory posture are also dependent upon displays being located a suitable distance from the eyes. When focused on infinity, or any object more than around 6 m distant, the lens of the eye is completely relaxed. To look at closer objects than this requires effort, both of the orbital muscles for convergence and of muscle within the eye itself for accommodation. In young people the processes of convergence and accommodation reach their limits or 'near points' at around 80 and 120 mm, respectively. (The latter increases dramatically with age as the lens of the eye stiffens; this also reduces the rate at which the eye can accommodate to different distances.) Visual work performed excessively close to the eyes is fatiguing and leads to 'eyestrain' – a poorly defined condition involving blurring of vision, headache and burning or 'gravelly' sensations around the eyes. As is the case with most criteria, there is no sharp cut-off point for minimum acceptable viewing distance and authorities differ in the figures they recommend. Figures as low as 350–400 mm are sometimes quoted, and indeed may be acceptable under certain circumstances. But for most practical purposes a minimum viewing distance of about 500 mm is probably desirable; and 750 mm or more may well be preferable (provided that the visual display is sufficiently bold to be read at this distance; see also Chapter 6). The VDU operators studied by Grandjean *et al.* (1984) adjusted their workstations to an average visual distance of 760 mm (settings ranged from

610 to 930 mm). The data of Brown and Schuam (1980) give an average preferred figure of 624 mm.

It is interesting to note that pain and spasm in the neck muscles (trapezius, sternomastoid, splenius, etc.) can lead to 'mechanical headache' – experienced in various parts of the head and face and not uncommonly around or 'behind' the eyes (Travell 1967, Dalassio 1980, Travell and Simons 1983). (Anatomists reading this will note the proximity of the proprioceptive supply of these muscles to the spinal nucleus of the trigeminal nerve.) The symptoms of mechanical headache and eyestrain are exceedingly similar (see also Chapter 8).

3.8 Working height

The height above the ground at which manual activities are performed by the standing person is a major determinant of that person's posture. If the working level is too high the shoulders and upper limbs will be raised, leading to fatigue and strain in the muscles of the shoulder region (trapezius, deltoid, levator scapulae, etc.). If any downward force is required in the task the upper limbs will be in a position of poor mechanical advantage for providing it. This problem may be avoided if the working level is lower. One commonly hears people talk of 'using their weight' or 'getting their weight on top of' the action. This is probably a misconception: what we really mean is that a vertical force may be exerted with minimal loading to the elbow and shoulder extensor muscles. A downward force, however exerted, can never exceed body weight (unless your feet are bolted to the floor), but in some positions the muscles of your arm may lack the strength to lift your feet off the ground.

If, however, the working level is too low the trunk, neck and head will be inclined forwards with consequent postural stress for the spine and its muscles. It may be presumed that somewhere between a working level that is too high and one that is too low there may be found a suitable compromise at which neither the shoulders nor the back are subjected to excessive postural stress.

It is important to distinguish between working height and work-surface height. The former may be substantially higher than the latter if hand tools or other equipment are being used in the task. In some cases the working level may actually be below the work surface – consider the task of washing up which, in the conventional kitchen, is performed in a recess set into the working surface (i.e. the sink).

The following recommendations concerning working height are widely quoted (see, e.g. Grandjean 1988, Pheasant 1987, 1991a,b):

- for manipulative tasks involving a moderate degree of both force and precision – 50–100 mm below elbow height;
- for delicate manipulative tasks (including writing) – 50–100 mm above elbow height (wrist support will generally be necessary);
- for heavy manipulative tasks (particularly if they involve downward pressure on the workpiece) – 100–250 mm below elbow height;
- for lifting and handling tasks – between knuckle height and elbow height (see also Chapter 8);
- for hand-operated controls (e.g. switches, levers, etc.) – between elbow height and shoulder height (see also Section 3.3).

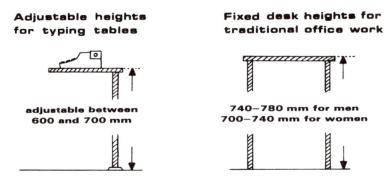

Figure 3.14   Recommended desktop heights for traditional office jobs. Left: range of adjustability for typing desks; right: desktop heights for reading and writing without typewriters. (From E. Grandjean, *Fitting the Task to the Man*, Taylor & Francis, 1988, 4th Edn, fig. 37, p. 42; reproduced with kind permission.)

These recommendations are summarized in Figure 3.14. For a discussion of how criteria of this kind may be applied to the dimensioning of workspaces and equipment, see Chapter 2.

3.9   Posture and strength

Studies in which strength is measured in different positions commonly show that the differences between conditions (i.e. between working postures) are greater than the differences between individuals. Strength is dependent on posture, first for reasons of physiology, and second for reasons of simple mechanics.

The function of a muscle is to exert tension between its points of bony attachment – and by doing so to exert a torque or moment about the joint (or joints) that the muscle crosses. The capacity of a muscle to exert tension is dependent upon its length, which in turn is dependent upon the position (i.e. angle) of the joint across which it acts. This defines the *angle–torque relationship* of the joint. In general, muscles are able to exert their greatest tension in their 'outer range', that is, at or near their position of maximum length. Thus we should expect actions of flexion to be strongest starting from positions of extension and vice versa – at least if we measure strength in terms of torque about a joint. (For a further discussion of these matters and of muscle function in general, see Pheasant 1991a.)

In analysing real-world problems involving the exertion of forces on external objects, however, we shall generally also have to take account of the mechanical advantage at which the muscles act through the bony levers of the limbs and trunk. Consider the action of the muscles that flex (i.e. bend) the elbow in the pulling actions shown in Figure 3.15. The torque about the elbow (T_e), required to exert a force F, is given by the equation

$$T_e = Fd \qquad\qquad (3.11)$$

where d is the perpendicular distance from the line of action of the force to the fulcrum of the elbow joint. Thus the amount of effort required from the person's muscles, to exert a given force, will be very much less in the position shown on the

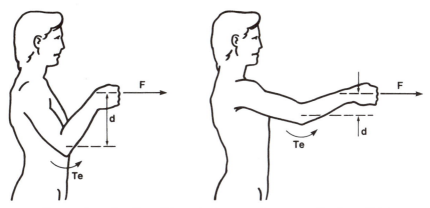

Figure 3.15 Torque about the elbow (T_e) required to exert the same pulling force (F) in two different postures.

right. In practice, leverage effects of this sort will in most cases be more important as determinants of the force you can exert in a particular situation than the angle–torque relationships of the joints involved.

As a general rule therefore, pushing or pulling actions are likely to be strongest (and thus to require least effort) when exerted along the line of an almost straight limb. Bicyclists know this – and adjust the saddle so that the leg is almost straight when the pedal is in its bottom position. Thus they minimize the perpendicular distance (averaged over the motion cycle) of the line of thrust from the knee joint, and thus minimize the amount of effort work required from the knee extensor muscles to do a given amount of work on the pedals.

For the same reason, lifting actions will be strongest (and also safest) when performed close to the body, that is, close to the fulcra of the articulations concerned, particularly those of the low back (see Chapter 8).

There are circumstances in which the force that a person can exert is limited by factors other than the capacity of his or her muscles, such as bodily support and stability, the deployment of body weight or the frictional resistance between the feet and the floor. These matters are discussed at length in Grieve and Pheasant (1982) and more briefly in Pheasant (1991a).

CHAPTER FOUR

Sitting and seating

4.1 Fundamentals of seating

The purpose of a seat is to provide stable bodily support in a posture that is:

(i) comfortable over a period of time;

(ii) physiologically satisfactory;

(iii) appropriate to the task or activity in question.

All seats are uncomfortable in the long run, but some seats become uncomfortable more rapidly than others, and in any particular seat, some people will be more uncomfortable than others. Comfort may also be influenced by the task or activity that the user is engaged in at the time. In other words, comfort (or more strictly the rate of onset of discomfort) will depend upon the interaction of *seat characteristics*, *user characteristics*, and *task characteristics* (Table 4.1).

   In matching the seat to the user, anthropometric factors are of major importance – but by no means uniquely so. An appropriate match between the dimensions of

Table 4.1 Determinants of sitting comfort.

Seat characteristics		User characteristics
▪ seat dimensions		▪ body dimensions
▪ seat angles		▪ body aches and pains
▪ seat profile		▪ circulation
▪ upholstery		▪ state of mind
	Task characteristics	
	▪ duration	
	▪ visual demands	
	▪ physical demands	
	— hands	
	— feet	
	▪ mental demands	

the seat and those of its users is necessary for comfort, but not sufficient. We shall return to the anthropometric aspects of seating in due course.

In general, a seat that is comfortable in the (relatively) long term will also be physiologically satisfactory. In one sense it is difficult to see how this could not be the case – given that the neural events that tell us that we are 'uncomfortable' may in physiological terms be regarded as warning signs of impending tissue damage. We might suppose therefore, that in the absence of such warnings, no damage is imminent. It may not be as simple as this, however. There are those who believe that extensive covert damage due to 'poor sitting posture' may occur in the absence of subjective discomfort. This is actually a very difficult argument to settle either way. To gain some further insight into these matters, we turn now to a consideration of the physiology and biomechanics of the sitting posture, with particular reference to the structure and function of the lumbar spine.

4.2   The spine in standing and sitting

The human vertebral column (backbone) consists of twenty-four movable bony vertebrae separated by deformable hydraulic pads of fibrocartilage known as intervertebral discs. (Up to 10% of people possess a greater or lesser number of vertebrae but these 'anomalies' seem to have little functional consequence.) The column is surmounted by the skull, and rests upon the sacrum which is firmly bound to the hip bones at the sacro-iliac joints. The vertebrae can be naturally grouped into seven cervical (in the neck), twelve thoracic (to which the ribs are attached) and five lumbar (in the small of the back, between the ribs and the pelvis). The spine is a flexible structure, the configuration of which is controlled by many muscles and ligaments (Figure 4.1).

In the upright standing position the well-formed human spine presents a sinuous curve when viewed in profile. The cervical region is concave (to the rear), the thoracic region convex and the lumbar region again concave. A concavity is sometimes

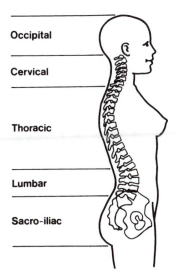

Figure 4.1   The well-formed human spine presents a sinuous curve when viewed in profile.

known as a 'lordosis' and a convexity as a 'kyphosis'. These are enclosed by the convexities of the occiput (back of the head) above with the sacro-iliac region and buttocks below; making five curves in all.

In the upright standing position, the pelvis is more or less vertical and the first lumbar vertebra and sacrum make angles of about 30° above and below the horizontal respectively (see Figure 4.2). Consider what happens when you sit down on a relatively high seat (such as a dining-room chair). You flex your knees through 90° and make another 90° angle between your thighs and trunk. Most of your weight is taken by the ischial tuberosities – two bony prominences which you can feel within the soft tissue of your buttocks if you sit on your hands. Part of the right angle between the thighs and trunk is achieved by flexion at the hip joint. After an angle of 60° is reached this movement is opposed, unless we are very flexible, by tension in the hamstring muscles (located in the backs of the thighs) hence we tend to complete the movement by a backward rotation of the pelvis of 30° or more – as shown on the left-hand side of Figure 4.3.

This backward rotation must be compensated by an equivalent degree of flexion in the lumbar spine – if the overall line of the trunk is to remain vertical. Hence in sitting down we tend to flatten out the concavity (lordosis) of the lumbar region.

In relaxed unsupported sitting, the lumbar spine may well be flexed close to the limit of its range of motion. In this position, the muscles will be relaxed, because the weight of the trunk will be supported by tension in passive structures such as ligaments. This is achieved, however, at the expense of a considerable degree of deformation of the intervertebral discs, the pads of fibrocartilage or 'gristle' which separate the bony vertebrae (see Figure 4.4). This is widely thought to be a bad

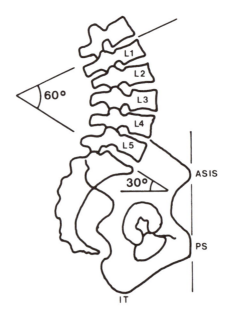

Figure 4.2   Typical orientation of the lumbar spine and pelvis in the standing position.
ASIS = anterior superior iliac spine; PS = pubic symphysis; IT = ischial tuberosity. (From S. Pheasant, *Ergonomics, Work and Health*, Macmillan, 1991, Figure 5.2, p. 102, reproduced with kind permission.)

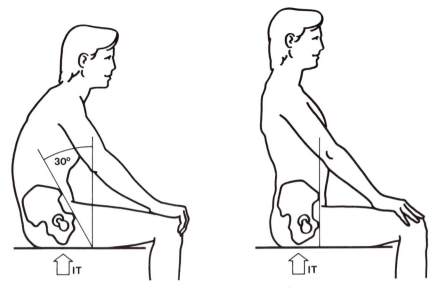

Figure 4.3   In relaxed sitting (left) the pelvis rotates backwards and the spine is flexed. To sit up straight (right) requires muscular exertion to pull the pelvis forward. The ischial tuberosities (IT) act as a fulcrum.

thing. (The reasons that this should be so are beyond the scope of the present discussion. Suffice it to say that in the author's view they are good ones.)

In order to 'sit up straight' and regain our lost lordosis we must make a muscular effort to overcome the tension in the hamstrings. (The effort probably comes from a muscle deep within the pelvis called iliopsoas.) We cannot merely relax the hamstrings since their tension is a passive one, caused by the stretching of tissue (just like an elastic band) rather than by actual muscular contraction. We shall probably

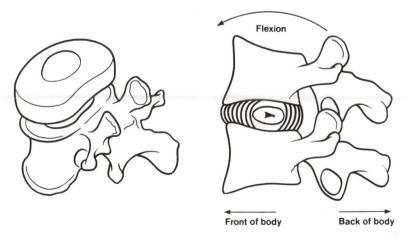

Figure 4.4   Left: lumbar vertebra surmounted by intervertebral disc, showing outer and inner parts. Right: deformation of the disc during flexion of the spine. (Redrawn from Kapanji (1974).)

also need to activate our back muscles to support the weight of our trunk. If prolonged, this static muscle loading may become a major source of postural discomfort, particularly in someone who has a pre-existing tendency to suffer from back
trouble.

In designing a seat therefore, the objective is to support the lumbar spine in its
neutral position (i.e. with a modest degree of lordosis) without the need for muscular
effort, thus allowing the user to adopt a position that is *both* physiologically satisfactory *and* comfortably relaxed. In general this will be achieved by:

(i) a semi-reclined sitting position (to the extent that this is permitted by the
     demands of the working task);

(ii) a seat that is neither lower nor deeper than necessary (see below);

(iii) a backrest that makes an obtuse angle to the seat surface (thus minimizing the
      need for hip flexion) and is contoured to the form of the user's lumbar spine.

The extent to which the backrest of the seat supports the weight of the trunk (and
thus reduces the mechanical loading on the lumbar spine) is a direct function of its
angle of inclination to the vertical. This may be predicted theoretically (as a simple
matter of cosines) – and it has been confirmed by Andersson *et al.* (1974) in a series
of experimental studies in which the hydrostatic pressure within the nucleus pul-
posus was measured directly using needle-mounted transducers. Andersson *et al.*
(1974) also found that for any given angle of backrest inclination the intra-discal
pressure was measurably less if the backrest was contoured to the form of the
lumbar spine (Figure 4.5).

Grandjean (1988) reported the results of a series of fitting trials using what he
called a 'sitting machine'. This was an adjustable test rig by means of which it was
possible to determine the preferred seat profiles of experimental subjects (or more
specifically, the profiles that minimized reported aches and pains during sitting). The
reported preferences of subjects who suffered from back trouble were much the same

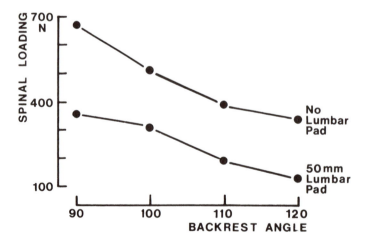

Figure 4.5 Spinal compression measured directly by needle-mounted transducer at seat back
angles from vertical (90°) to reclined (120°), with and without a pad in the lumbar region. (Data
from Andersson *et al.* (1974). From S. Pheasant, *Ergonomics, Work and Health*, Macmillan, 1991,
Figure 11.3, reproduced with kind permission.)

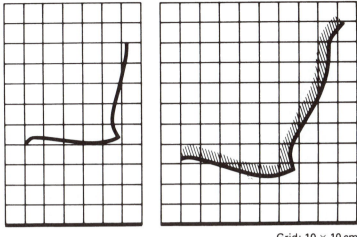

Grid: 10 × 10 cm

Figure 4.6 Seat profiles of a multipurpose chair (left) and an easy chair (right), both of which caused a minimum of subjective complaints. (From E. Grandjean, *Fitting the Task to the Man*, Taylor & Francis, 1988, 4th Edn, Figure, 52, p. 60, reproduced with kind permission.)

as those of people who did not. Figure 4.6 shows the overall preferred profiles (for both groups of subject) for a 'multi-purpose' chair and an 'easy chair'.

Andersson's pressure measurements and Grandjean's fitting trials confirm therefore that a seat that enables the user to adopt a semi-reclined position and has a backrest that is contoured to the shape of the lumbar spine will *both* minimize the mechanical loading on the lumbar spine *and* maximize the overall levels of reported comfort (both for users who suffer from back trouble and for those who do not).

A problem arises, however, in tasks such as writing – which entail forward leaning and in which the support of the backrest will tend to be lost. The backrest remains important in these activities, however, during rest pauses. Grandjean (1988) describes a study of office workers using time-lapse photography, which showed them to be in contact with the backrest for 42% of the time.

4.2.1   Forward tilt seating

In recent years a radical new approach to seat design has been proposed. Mandal (1976, 1981) argued that the seat surface should slope forwards, hence diminishing the need for hip flexion (particularly in tasks such as typing and writing) and encouraging lumbar lordosis. A number of seat designs now incorporate a tilt mechanism (see Figure 4.7). The disadvantage of such a design is that if you sit on the chair without thinking, you will tend to exert a backward thrust with the feet in order to stay in the seat – this is a particular problem if the chair is on castors. High-friction upholstery is not really an answer since female attire (in particular) generally provides a low-friction interface between the outer and inner garments – women, therefore, tend to slide out of their skirts. Experience suggests that balancing correctly on the forward slope seat is a skill that needs to be learned. According to Mandal (1981), users may take 1–2 weeks to get used to such chairs.

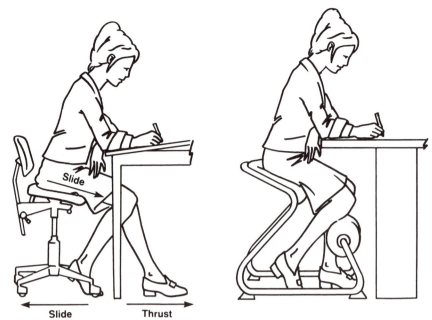

Figure 4.7 Two radical approaches to seat design: the forward tilting seat (left) and the kneeling chair (right).

These difficulties should be overcome with the 'kneeling chair' which provides a seat sloping forward at some 30° to the horizontal, combined with a padded support for the knees. Brunswic (1981) evaluated these concepts. She found that:

(i) when seat angle and knee angle were independently varied, forward tilted seat positions did not result in a lumbar posture that was significantly different from that obtained with a horizontal seat and the knees at right angles;

(ii) the lumbar posture of subjects using a kneeling chair in writing and typing tasks was not significantly better than when they used a conventional office chair.

Drury and Francher (1985) evaluated a kneeling chair by means of a user trial of considerable sophistication. Subjects were typing or operating a computer terminal. They concluded that the comfort was 'no better than conventional chairs and could be worse than well-designed office chairs'. The principal complaints were difficulties of access–egress, pressure on the shins, and discomfort in the knee region (due presumably to impairment of circulation resulting from the acute angle). There was 'little or no decrease in back discomfort' and, in spite of training, subjects 'often slumped forward to give a kyphotic spine'. Presumably they did this to rest their back muscles by 'hanging on their ligaments'.

The subjects in this trial all had 'normal' backs. Atherton *et al.* (1982) tested a kneeling chair, along with a number of conventional office chairs, on a group of subjects all of whom had musculoskeletal problems of one sort or another and half of whom had bad backs. The kneeling chair came about half-way down the list in their rank order of preferences.

The kneeling chair has two obvious disadvantages: standing up and sitting down is necessarily difficult; and it fixes the lower limbs with the knees in a position of flexion well past the mid-range. This reduces the possibilities for fidgeting and changes of posture (except by a backwards and forwards rocking of the pelvis). Since the chair (in its basic form) has no back, the loading on the lumbar spine cannot be any lower than it is in upright standing; whereas, when a person leans back fully on the backrest of a conventional seat, the loading on the lumbar spine may be very much less than in standing (Andersson *et al.* 1974, Pheasant 1991a).

Overall, such scientific studies as have been done of the matter, do not suggest that the kneeling chair offers any particular material advantages, as compared with a well-designed chair of the conventional sort, either with regard to sitting in general, or with regard to office use in particular. Having said this, however, one must also add that on the basis of clinical experience it is quite clear that *some* people who suffer from back trouble find the kneeling chair helpful. One's impression is that these people are only a minority of back-pain patients as a whole (and possibly a fairly small minority) – but experience suggests that those back-pain patients who like the kneeling chair often like it very much indeed. It would be interesting to know why: perhaps these are patients who have an unusually poor tolerance of spinal flexion.

4.3 Anthropometric aspects of seat design

(The following seat dimensions are shown in Figure 4.8.)

4.3.1 Seat height (H)

As the height of the seat increases, beyond the popliteal height of the user, pressure will be felt on the underside of the thighs. The resulting reduction of circulation to the lower extremities may lead to 'pins and needles', swollen feet and considerable discomfort. As the height decreases the user will (a) tend to flex the spine more (due to the need to achieve an acute angle between thigh and trunk); (b) experience greater problems in standing up and sitting down, due to the distance through which his centre of gravity must move; and (c) require greater leg room. In general, therefore, the optimal seat height for many purposes is close to the popliteal height, and where this cannot be achieved a seat that is too low is preferable to one that is too high. For many purposes, therefore, the 5th %ile female popliteal height (400

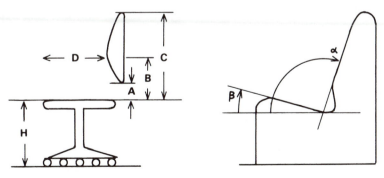

Figure 4.8   Seat dimensions.

mm shod) represents the best compromise. If it is necessary to make a seat higher than this (e.g. to match a desk or because of limited leg room), the ill effects may be mitigated by shortening the seat and rounding off its front edge in order to minimize the under-thigh pressure. It is of overriding importance that the height of a seat should be appropriate to that of its associated desk or table.

4.3.2   Seat depth (D)

If the depth is increased beyond the buttock–popliteal length (5th %ile woman = 435 mm), the user will not be able to engage the backrest effectively without unacceptable pressure on the backs of the knees. Furthermore, the deeper the seat the greater the problems of standing up and sitting down. The lower limit of seat depth is less easy to define. As little as 300 mm will still support the ischial tuberosities and may well be satisfactory in some circumstances. Tall people some-times complain that the seats of easy chairs are too short – an inadequate backrest may well be to blame (see below).

4.3.3   Seat width

For purposes of support a width that is some 25 mm less on either side than the maximum breadth of the hips is all that is required – hence 350 mm will be ade-quate. However, clearance between armrests must be adequate for the largest user. The hip breadth of the 95th %ile woman unclothed is 435 mm. In practice, allowing for clothing and leeway, a minimum of 500 mm is required. (In some cases elbow–elbow breadth, as given in Table 2.3, is more relevant: 95th %ile clothed man = 550 mm.)

4.3.4   Backrest dimensions

The higher the backrest, the more effective it will be in supporting the weight of the trunk. This is always desirable but in some circumstances other requirements such as the mobility of the shoulders may be more important. We may distinguish three varieties of backrest, each of which may be appropriate under certain circumstances: the low-level backrest; the medium-level backrest and the high-level backrest.

The *low-level backrest* provides support for the lumbar and lower thoracic region only and finishes below the level of the shoulder blades, thus allowing freedom of

Table 4.2   Typical recommendations concerning backrest dimensions of work chairs.

	BS 5940	HSE (1991)
Seat surface to bottom of backrest (A)		100–200
Seat surface to mid-point of lumbar curve (B)		
— fixed height backrest	210 ± 15	
— adjustable height backrest	170–250	170–300
Vertical height of backrest (C–A)		200–550

Note: All dimensions in mm.

movement for the shoulders and arms. Old-fashioned typists' chairs generally had low-level backrests, as do many general purpose stacking chairs. To support the lower back and leave the shoulder regions free, an overall backrest height (C) of about 400 mm is required.

The *medium-level backrest* also supports the upper back and shoulder regions. Most modern office chairs fall into this category, as do many 'occasional' chairs, auditorium seats, etc. For support to mid-thoracic level an overall backrest height of about 500 mm is required and for full shoulder support about 650 mm (95th %ile male values rounded up). A figure of 500 mm is often quoted for office chairs (see Chapter 6).

The *high-level backrest* gives full head and neck support – for the 95th %ile man an overall backrest height of 900 mm is required.

Whatever its height, it will generally be preferable and sometimes essential for the backrest to be contoured to the shape of the spine, and in particular to give 'positive support' to the lumbar region in the form of a convexity or pad (see above).

To achieve this end, the backrest should support you in the same place as you would support yourself with your hands to ease an aching back. To use the lumbar support to its full advantage, it is also necessary to provide clearance for the buttocks – so in some kinds of chair (including work chairs) it may be appropriate to leave a gap between the seat surface and the bottom edge of the backrest. For work chairs an adjustable backrest is usually desirable and in some contexts essential. Some typical recommendations are summarized in Table 4.2.

A medium- or high-level backrest should be flat or slightly concave above the level of the lumbar pad. But the contouring of the backrest should in no cases be excessive in fact a curve that is too pronounced is probably worse than no curve at all. Andersson *et al.* (1974) found that a lumbar pad that protrudes 40 mm from the main plane of the backrest at its maximum point will support the back in a position that approximates to that of normal standing.

4.3.5 Backrest angle or 'rake' (α)

As the backrest angle increases, a greater proportion of the weight of the trunk is supported – hence the compressive force between the trunk and pelvis is diminished (and with it intradiscal pressure). Furthermore, increasing the angle between trunk and thighs improves lordosis. However, the horizontal component of the compressive force increases. This will tend to drive the buttocks forward out of the seat unless counteracted by (*a*) an adequate seat tilt; (*b*) high-friction upholstery; or (*c*) muscular effort from the subject. Increased rake also leads to increased difficulty in the stand-up/sit-down action.

The interaction of these factors, together with a consideration of task demands, will determine the optimal rake which will commonly be between 100° and 110°. A pronounced rake (e.g. greater than 110°) is not compatible with a low- or medium-level backrest since the upper parts of the body become highly unstable.

4.3.6 Seat angle or 'tilt' (β)

A positive seat angle helps the user to maintain good contact with the backrest and helps to counteract any tendency to slide out of the seat. Excessive tilt reduces

hip/trunk angle and ease of standing up and sitting down. For most purposes 5–10°
is a suitable compromise. (See also Chapter 6.)

4.3.7   Armrests

Armrests may give additional postural support and be an aid to standing up and
sitting down. Armrests should support the fleshy part of the forearm, but unless very
well padded they should not engage the bony parts of the elbow where the highly
sensitive ulnar nerve is near the surface; a gap of perhaps 100 mm between the
armrest and the seat back may, therefore, be desirable. If the chair is to be used with
a table the armrest should not limit access, since the armrest should not, in these
circumstances, extend more than 350 mm in front of the seat back. An elbow rest
that is somewhat lower than sitting elbow height is probably preferable to one that
is higher, if a relaxed posture is to be achieved. An elbow rest 200–250 mm above
the seat surface is generally considered suitable.

4.3.8   Leg room

In a variety of sitting workstations the provision of adequate lateral, vertical, and
forward leg room is essential if the user is to adopt a satisfactory posture.

Lateral leg room

Lateral leg room (e.g. the 'kneehole' of a desk) must give clearance for the thighs and
knees. In a relaxed position they are somewhat separated: BS 5940 quotes a
minimum width of 580 mm.

Vertical leg room

Requirements will, in some circumstances, be determined by the knee height of a tall
user (95th %ile shod man = 620 mm). Alternatively, thigh clearance above the
highest seat position may be more relevant – adding the 95th %ile male popliteal
height and thigh thickness gives a figure of 700 mm (BS 5940 quotes a minimum of
650 mm for a general purpose desk).

Forward leg room

This is rather more difficult to calculate. At knee level clearance is determined by
buttock–knee length from the back of a fixed seat (95th %ile male = 645 mm). If the
seat is movable we may suppose that the user's abdomen will be in contact with the
table's edge (although, in practice, most people will choose to sit further back than
this). In this case clearance is determined by buttock–knee length minus abdominal
depth, which will be around 425 mm for a male who is a 95th %ile in the former
and a 5th %ile in the latter. At floor level an additional 150 mm clearance for the
feet gives a figure of 795 mm from the seat back or 575 mm from the table's edge.
All of these figures are based on the assumption of a 95th %ile male sitting on a seat
that is adjusted to approximately his own popliteal height, with his lower legs verti-

cal. If the seat height is in fact lower than this he will certainly wish to stretch his legs forward. A rigorous calculation of the 95th %ile clearance requirements in these circumstances would be complex but an approximate value may be derived as follows.

Consider a person of buttock–popliteal length B, popliteal height P, and foot length F sitting on a seat of height H (as shown in Figure 4.9). He stretches out his legs so that his popliteal region is level with the seat surface (i.e. his thighs are approximately horizontal). The total horizontal distance between buttocks and toes (D) is approximated by

$$D = B + \sqrt{P^2 - H^2} + F \qquad (4.1)$$

(ignoring the effects of ankle flexion). Hence, in the extreme case of a male who is a 95th %ile in the above dimensions, sitting on a seat that is 400 mm in height requires a total floor level clearance of around 1190 mm from the seat back or 970 mm from the table edge (if he is also a 5th %ile in abdominal depth). Such a figure is needlessly generous for most purposes; most ergonomics sources quote a minimum clearance value of between 600 and 700 mm from the table edge. (BS 5940 quotes minima of 450 mm at the underside of the desk top and 600 mm at floor level and for 150 mm above.)

4.3.9 Seat surface

The purpose of shaping or padding the seat surface is to provide an appropriate distribution of pressure beneath the buttocks. The consensus of ergonomic opinion suggests the following:

(i) the seat surface should be more or less plane rather than shaped, although a rounded front edge is highly desirable;

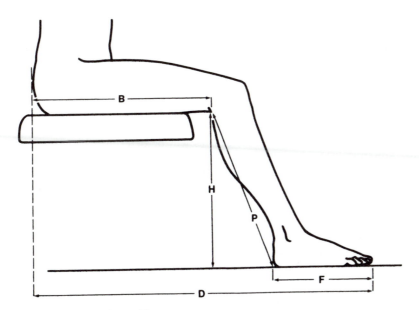

Figure 4.9 Calculation of forward leg room.

Table 4.3   Sitting in a row.

	Width required (mm)		
Number of persons	Mean	SD	95th %ile
1	480	28	526
2	960	40	1026
3	1440	48	1519
4	1920	56	2012

(ii) upholstery should be 'firm' rather than 'soft' (it is sometimes said that a heavy
user should not deform it by more than 25 mm);

(iii) covering materials should be porous for ventilation and rough to aid stability.

The traditional wooden 'Windsor' chair can be surprisingly comfortable in spite of
its total absence of upholstery. Its basic form was probably developed by the craftsmen of the Chiltern beechwoods sometime around the beginning of the eighteenth
century. A critical feature seems to be the subtle contouring of the seat known as its
'bottoming'. This was hand carved, using first an adze then a variety of shapers, by a
man known as the 'bottomer', whose specialized trade was considered the most
skilled of all the activities that contributed to the chair-making process. He worked
by eye without recourse to measurements; contemporary machine-made versions
are said to be less satisfactory.

4.3.10   Seats for more than one

When considering benches and other seats in which users sit in a row, it is necessary
to bear in mind that the breadth of a 95th %ile couple is less than twice that of a
95th %ile individual. (The chance of two people, each 95th %ile or more, meeting at
random on a bench is only 1 in 400.) In general, n people sitting in a row have a
mean breadth of nm and a standard deviation of $s\sqrt{n}$, where m and s are the parameters of the relevant body breadth – which will usually be that of the shoulders.
Table 4.3 gives values based on male data and including a clothing correction of 15
mm.

  However, if the row of seats is divided by armrests the problem is more complex.
Assume each user sits in the centre of his seating unit – a little reflection will tell us
that the minimum separation of seat centres will be determined by the distribution
of pairs of half-shoulder breadths: 480 [40] mm; 95th %ile = 545 mm. Since in the
presence of armrests the minimum seat breadth is 500 mm (see above), and an
armrest cannot reasonably be less than 100 mm wide, 600 mm between seat centres
will satisfy all criteria.

4.4   The easy chair and its relatives

The function of an easy chair is to support the body during periods of rest and
relaxation. If not actually dozing or engaged in peaceful contemplation, the user
may be reading, watching television or in conversation. The form of the chair

follows naturally from these functions and from the considerations of the previous section.

Grandjean (1973) recommends a seat tilt (β) of 20–26° and an angle between seat and backrest of 105–110°. This gives a backrest rake (α) of as much as 136°, which is really only suitable for 'resting' and requires a degree of agility for standing up and sitting down. Le Carpentier (1969) found a tilt of 10° with a rake of 120° to be suitable for both reading and watching television. The present author inclines to the latter view with the caveat that for elderly users a rake of more than 110° may cause problems. Difficulties of standing up and sitting down will be reduced if the space beneath the front of the chair is unimpeded, allowing the user to place his feet beneath his centre of gravity, hence achieving a more vigorous upward thrust and a more controlled descent.

A high-level backrest is virtually essential to the proper role of an easy chair in providing support for the trunk. Its shaping is something of a challenge. It is possible to design a gentle lumbar curve that will suit most users, but an equivalent pad for the neck and occiput is more problematic. Ideally, this should give you similar support to the natural action of clasping your hands behind your head. A sensible way of achieving this is to incline the upper part of the backrest forwards from the main rake by around 10° and to provide a movable cushion. (This solution has been adopted on certain British Rail seats but, unfortunately, the range of adjustment of the cushion is not quite adequate for the shorter person.)

The fundamental problems of designing an easy chair had essentially been solved by around 1680, as the collection of almost any English country house will testify. Ergonomic research has merely confirmed the intuitions of the designers of the past. However, the present-day furniture showroom typically presents a range of styles that, in ergonomic terms, are rarely better than just adequate and not infrequently fall short on numerous criteria. There are, of course, exceptions but these are commonly either reworkings of traditional types (such as the ever popular 'William and Mary') or else chairs that are described as 'orthopaedic' and sold more as 'aids' than as the furnishings of a stylish home.

The most common failings in the contemporary armchair are a seat that is too deep and a backrest that is too low. One may suppose that this is due to an attempt to make the seat and back equal in length in the interests of visual symmetry (like the Mies Van der Rohe 'Barcelona' chair of 1929) or to an even more misguided attempt to fit the entire chair into a cubic outline (like Frank Lloyd Wright's 'Cube' chair of 1985 or 'Le Grand Confort' by Le Corbusier and Charlotte Perriaud of 1928–1929). Combined with the weighty stylistic influence of these modern masters is a marketing need to incorporate the armchair into a three-piece suite (or some other combination). With the exception of a few historical types, such as the William and Mary 'love seat', high-backed settees are virtually unknown. In reality, as anthropometric data quite clearly show, the backrest height needs to be around twice the seat depth if an easy chair is to perform its proper function.

Tall people sometimes complain of seats being insufficiently deep (i.e. too short from front to back). Observation suggests that on engaging the backrest and finding that it only reaches mid-shoulder level, they move down into the seat in an attempt to gain head support. As a result their buttocks slide forwards until they are in danger of dropping off the front of the seat. (This also leads to the flexed position which is physiologically least satisfactory.) Hence the problem stems from an inadequate backrest rather than a seat that is not deep enough.

A common misconception, held by designers and consumers alike, is to equate
depth and softness of upholstery with comfort. The luxurious sensation of sinking
into a deep over-stuffed sofa is indicative of an absence of the support necessary for
long-term comfort in the sitting position. In functional terms, we are now dealing
with something more amorphous than a seat *per se*, it is in fact an object for sprawling or reclining on, rather than for conventionally sitting on. Structurally, however,
the object retains the form of a seat. A seat supports its user in a sitting position and
a bed supports him in recumbent position – but there are a whole variety of intermediate sprawling postures which can be perfectly satisfactory, especially when, supported by mounds of cushions, one has the opportunity for frequent postural
changes. Taken to its logical conclusion the concept of 'amorphous furniture', which
does not dictate any posture in particular, leads to items such as the 'sag bag' – a
sack full of polystyrene beads, which enjoyed a brief vogue among young home-
makers a decade or so ago. A whole family of all but extinct furniture types, which
generically we could call couches, are essentially designed for sprawling – notable
members of this family are the 'day bed' mentioned in Shakespeare (*Twelfth Night*,
II.v) and the chaise-longue. A steeply raked easy chair can double as a couch when
used in conjunction with a footstool – as in the ergonomically excellent Charles
Eames lounge chair and ottoman of 1956 (Figure 4.10). The three-piece suite aims to
serve for both sitting and sprawling. It commonly does both tolerably but excels at
neither. There is considerable scope for design innovation in changing this state of
affairs.

Figure 4.10   The Charles Eames lounge chair and ottoman (1956) give good support in a wide
variety of postures.

CHAPTER FIVE

Hands and handles

5.1 Anthropometry of the hand

Table 5.1 gives anthropometric data for the adult hand, gathered together from a number of sources. It may be assumed that these figures are for a population of British adults equivalent to that of the 'standard reference population' as described in Table 2.3. The dimensions are illustrated in Figure 5.1. Hand length and hand breadth (1 and 12) are quoted directly from Table 2.3; dimensions 2–11, 13 and 15 are quoted from Kember *et al.* (1981); dimensions 16, 17 and 19 are quoted from Gooderson *et al.* (1982); dimension 20 is quoted directly from Davies *et al.* (1980) for women and estimated by scaling for men, and dimension 18 is scaled down from Garret (1971).

5.2 Anatomical terminology

Standard anatomical terms that are used to describe the position and movements of the forearm, wrist and hand are illustrated in Figure 5.2. The movements of flexion, extension and radial and ulnar deviation occur at the wrist joint complex – that is at the 'true' wrist (radiocarpal) joint and at the various articulations which are present between the eight small bones of the wrist (intercarpal joints). Ulnar deviation is sometimes also known as 'adduction' of the wrist and radial deviation as 'abduction' – but the terms are confusing and are best avoided.

The forearm has two long bones – the radius and ulna – which run from the elbow to the wrist and articulate with each other at their top and bottom ends. When the hand is in its 'palms up' or *supine* position, these two bones are parallel. (The radius is on the thumb side; the ulna is on the little finger side.) As the hand is turned into the 'palms down' or *prone* position, the lower end of the radius rotates about the axis of the ulna and the shafts of the two bones cross. Note then that the movements of *pronation* and *supination* occur at the two articulations between the radius and ulna rather than at the wrist as such. In practice, however, the natural

Table 5.1 Anthropometric estimates for the hand (all dimensions in mm).

Dimension	Men 5th %ile	Men 50th %ile	Men 95th %ile	SD	Women 5th %ile	Women 50th %ile	Women 95th %ile	SD
1. Hand length	173	189	205	10	159	174	189	9
2. Palm length	98	107	116	6	89	97	105	5
3. Thumb length	44	51	58	4	40	47	53	4
4. Index finger length	64	72	79	5	60	67	74	4
5. Middle finger length	76	83	90	5	69	77	84	5
6. Ring finger length	65	72	80	4	59	66	73	4
7. Little finger length	48	55	63	4	43	50	57	4
8. Thumb breadth (IPJ)[a]	20	23	26	2	17	19	21	2
9. Thumb thickness (IPJ)	19	22	24	2	15	18	20	2
10. Index finger breadth (PIPJ)[b]	19	21	23	1	16	18	20	1
11. Index finger thickness (PIPJ)	17	19	21	1	14	16	18	1
12. Hand breadth (metacarpal)	78	87	95	5	69	76	83	4
13. Hand breadth (across thumb)	97	105	114	5	84	92	99	5
14. Hand breadth (minimum)[c]	71	81	91	6	63	71	79	5
15. Hand thickness (metacarpal)	27	33	38	3	24	28	33	3
16. Hand thickness (including thumb)	44	51	58	4	40	45	50	3
17. Maximum grip diameter[d]	45	52	59	4	43	48	53	3
18. Maximum spread	178	206	234	17	165	190	215	15
19. Maximum functional spread[e]	122	142	162	12	109	127	145	11
20. Minimum square access[f]	56	66	76	6	50	58	67	5

Notes:

[a] IPJ is the interphalangeal joint, i.e. the articulations between the two segments of the thumb;

[b] PIPJ is the proximal interphalangeal joint, i.e. the finger articulation nearest to the hand;

[c] as for dimension 12, except that the palm is contracted to make it as narrow as possible;

[d] measured by sliding the hand down a graduated cone until the thumb and middle fingers only just touch;

[e] measured by gripping a flat wooden wedge with the tip end segments of the thumb and ring fingers;

[f] the side of the smallest equal aperture through which the hand will pass.

hand movements we use in everyday life very often entail actions of pronation and supination in combination with movements occurring at the wrist.

Place your hand in your lap in a palms-up (supine) position and allow it to relax completely. It will naturally adopt what anatomists call the *position of rest* (Figure 5.3) – in which the fingers and thumb are slightly flexed. This is the position in which the resting tension in the muscles that respectively flex (i.e. bend) and extend (i.e. straighten) the fingers are in equilibrium.

Anatomists have made a number of attempts to classify the infinite variety of actions of which the human hand is capable. The most basic distinction is between gripping (or 'prehensile') actions of various kinds, and non-gripping actions (such as poking, pressing, stroking, slapping, etc). In a gripping action the hand forms a *closed kinetic chain* which encompasses the object in question; in a non-gripping action the hand is used in an 'open chain' configuration. A few common everyday actions fall between these two categories, in that the kinetic chain of the hand is on

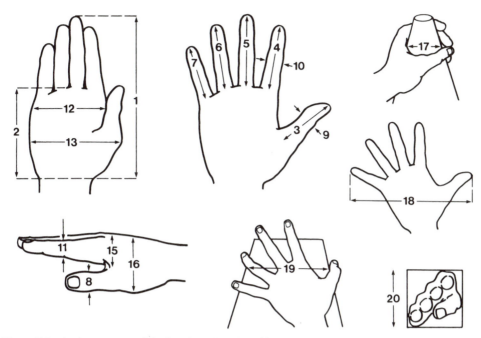

Figure 5.1 Anthropometry of the hand, as given in Table 5.1.

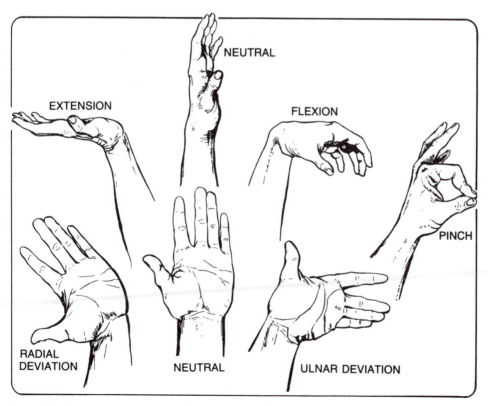

Figure 5.2 Hand and wrist postures. (From V. Putz-Anderson, *Cumulative Trauma Disorders*, Taylor & Francis, 1988, fig. 15, p. 54, reproduced with kind permission.)

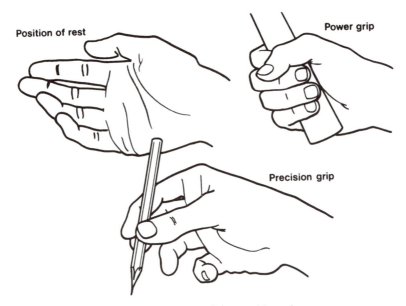

Figure 5.3   The position of rest, the power grip, and the precision grip.

the point of closing – for example, the hooking action that we use to carry a heavy suitcase and the action by which we scoop up a handful of small objects.

In a classic and widely quoted paper on the subject, Napier (1956) divided gripping actions into two main categories (see Figure 5.3):

(i) *power grips*, in which the fingers (and sometimes the thumb) are used to clamp the object against the palm;

(ii) *precision grips*, in which the object is manipulated between the tips (pads or sides) of the fingers and thumb.

Note that both entail a closed kinetic chain.

Although this classification will take us quite a long way in understanding hand function, it is something of an oversimplification. In the basic power grip shown in Figure 5.3 the thumb wraps around the back of the fingers to provide extra stability and gripping force. As the need for precision increases, however, the thumb moves along the shaft of the tool handle – providing extra control and the possibility of both power gripping and precision manipulation as the situation may demand. For a further discussion see Pheasant (1991a).

5.3   Fundamentals of handle design

The purpose of a handle is to facilitate the transmission of force from the musculo-skeletal system of the *user*; to the *tool* or object he is using; in the performance of the *task* or purpose for which he is using it (see Figure 1.1). As a general rule we can say that to optimize force transmission is to optimize handle design.

The following guidelines stem more from common sense than scientific investigation. They are commonly violated.

(i) Force is exerted most effectively when hand and handle interact in compression rather than shear. Hence, it is better to exert a thrust perpendicular to the axis of a cylindrical handle than along the axis (F_b in Figure 5.4 rather than F_a). If the latter is necessary a knob on the end will give extra purchase.

(ii) All sharp edges or other surface features, which cause pressure hot spots when gripped, should be eliminated. These include:

   (a) 'finger shaping' (unless designed with anthropometric factors in mind);

   (b) the ends of tools such as pliers, which may dig into the palm (if the handle is short);

   (c) the edges of flat or raised surfaces, e.g. for the application of labels, logos, etc.;

   (d) 'pinch points' between moving parts such as triggers, etc.

(iii) Handles of circular cross-section (and appropriate diameter, e.g. 30–50 mm) will be most comfortable to grip since there will be no possibility of hot spots – but they may not provide adequate purchase. Rectangular or polyhedral sections will give greater purchase but will be less comfortable. In general, wherever two planes meet (within the area that engages the hand) the edges should be rounded; there are no exact figures but a minimum radius of curvature of about 25 mm seems reasonable.

(iv) Surface quality should neither be so smooth as to be slippery nor be so rough as to be abrasive. The frictional properties of the 'hand/handle interface' are complex since the skin is both visco-elastically deformable and lubricated. Heavily varnished wooden handles give a better purchase than metal or plastic of similar smoothness. The explanation is possibly in their resilience (elastic compliance). Rubber is similar but becomes 'tacky'. The subject is worthy of more extensive investigation.

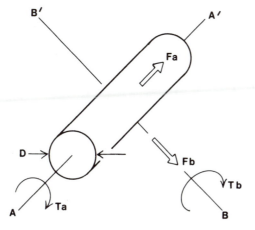

Figure 5.4   The cylindrical handle showing the long axis A–A' and the perpendicular axis B–B'.

(v) If part of the hand is to pass through an aperture (as in a suitcase or teacup) adequate clearance must be given. It is remarkable how often this perfectly obvious design principle is violated. The following spaces will accommodate virtually all users with a slight leeway:

- For the palm, as far as the web of the thumb (as in the handle of a suitcase), allow a rectangle 115 mm × 50 mm;
- For a finger or thumb, a circle 35 mm in diameter will allow insertion, rotation and extraction.

5.4 Biomechanics of tool design

5.4.1 Gripping and squeezing

An important group of cutting and crushing tools, from pliers and wire-cutters to nut-crackers and secateurs, are operated by a forceful squeezing action across two pivoting arms. The fingers curl around one arm and the heel of the palm butts against the other. The effective cutting/crushing force is determined by the mechanical advantage of the tool and the user's grip strength. The latter is determined inter alia by the distance across the two arms – as shown in Figure 5.5. The optimal handle separation is 45–55 mm for both men and women.

5.4.2 Gripping and turning

Consider a cylindrical handle as shown in Figure 5.4. It may be gripped and turned about its own axis $A–A'$ or about a perpendicular axis $B–B'$. Screwdrivers employ the former action; T-wrenches the latter.

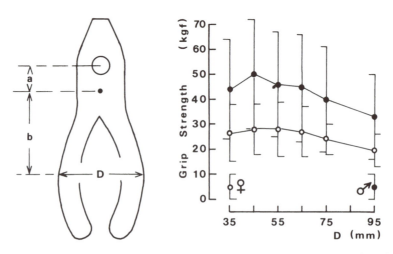

Figure 5.5   Grip strength (G) as a function of the handle size (D). Vertical lines are 5th–95th %ile values in samples of 22 men and 22 women. The tool is a lever of the first class, the mechanical advantage $= b/a$. Hence, the effective cutting or crushing force $= GB/A$. (Data from Pheasant and Scriven 1983.)

When the handle is employed as a T-wrench, the available torque (T_b) is, within reasonable limits, independent of the design of the handle.

When the handle is employed as a screwdriver (rotating about axis A–A') the strength of the action is no longer determined by the user's capacity to generate torque but by the ability to transmit it across the hand–handle interface. It is, therefore, strongly dependent upon handle design. Torque about axis A–A' is exerted by a shearing (frictional) action on the cylinder's surface, hence,

$$T_a = G\mu D \tag{5.1}$$

where G is the net compressive force (i.e. grip), D is the diameter of the cylinder, and μ is the coefficient of limiting friction between the hand and the handle. For any handle of circular cross-section (i.e. cylinder, sphere or disc), T_a will increase with diameter. We should also expect the grip strength (G) to be dependent upon diameter (the optimum value of which can only be determined empirically). Figure 5.6 summarizes the results of such an experiment. Few 'real' handles are actually circular in cross-section – but quite substantial irregularities in shape seem to make surprisingly little difference. Hence, commercially available screwdrivers (London pattern, cabinet makers', engineers, etc.) perform no better in these tests than do knurled steel cylinders of equivalent diameter (Pheasant and O'Neill 1975, Pheasant and Scriven 1983). Subsequent (unpublished) experiments have shown that the same is true for a variety of devices such as taps and doorknobs. However, torques exerted about axis B–B', as in using T-shaped or L-shaped devices, are very much

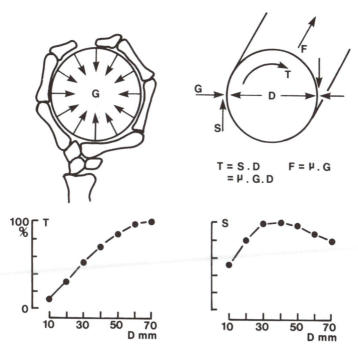

Figure 5.6   The mechanics of the gripping and turning action, using a cylindrical handle. Note that torque (T) is the greatest on the 70 mm handle, whereas both shear ($S = T/D = \mu\,G$) and thrust (F, not plotted), are greatest on handles in the 30–50 mm range. (From S. Pheasant, *Ergonomics, Work and Health*, Macmillan, 1991, fig. 14.5, p. 267, reproduced with kind permission.)

greater. (The torque that may be exerted using a typical L-shaped lever-type door handle is in the order of twice that available from any cylinder, sphere, or disc turning about its own axis.)

The strength of a thrusting action along axis A–A' is given by

$$F_a = G\mu \tag{5.2}$$

The diameter is only relevant as a determinant of G. Hence, we find that the optimal diameter for axial thrusts is somewhat less than that for turning actions. Data are summarized in Table 5.2. It is also worth noting that the maximal hand–handle contact area occurs on handles 50–60 mm in diameter – which will, therefore, minimize the surface stress to the skin (Pheasant and O'Neill 1975).

5.4.3 The neutral position of the wrist

Grip strength is greatest when the wrist is in its neutral position – falling off progressively as the wrist moves away from the neutral position in any direction (i.e. flexion, extension, radial deviation, ulnar deviation). The strength of grip is least when the wrist is flexed. This is because when the wrist is flexed, the finger flexors (which are the prime movers in the gripping action) are shortened and their capacity to generate tension is thus diminished (see Section 3.9).

For this reason alone, it would seem desirable that the handles of tools should be designed a such a way that when the tool is in use the wrist should remain as close as possible to the neutral position, since the less the strength of the gripping action in a given position, the harder the muscles will have to work to maintain a given level of gripping force. There are also other reasons.

The tendons of the various forearm muscles that act on the fingers and hand run around a variety of bony and ligamentous 'pulleys' where they cross the line of the wrist joint. When the wrist is in a non-neutral position, the mechanical loading on the tendons at these points of contact will be increased. (This is a matter of basic mechanics.) This increase in loading may lead to an increase in the 'wear and tear'

Table 5.2 Handle sizes that allow the greatest force/ torque in operation.

Pivoting tools	
distance across arms (mm)	45–55
Handles of circular cross-section	diameter (mm)
cylinders	
axial thrust (F_a)	30–50
axial rotation (T_a)	50–65
spheres	
axial rotation	65–75
discs	
axial rotation	90–130

Note: For cylindrical handles used to exert force or torque perpendicular to the axis (F_b, T_b) the diameter is not critical; a diameter of 30–50 mm is suitable.

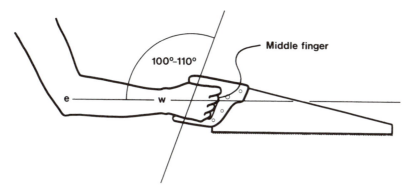

Figure 5.7 The neutral position of the wrist is preserved if the axis of grip makes an angle of 100–110° with the axis of the forearm.

on the tendons which the working task entails, and to the development of conditions like tenosynovitis, carpal tunnel syndrome, and other work-related musculo-skeletal disorders attributable to over-use (see Chapter 8).

When the wrist is in its neutral position, the long axis of a cylindrical handle that is held firmly in the hand, makes an angle of 100–110° to the axis of the forearm (Figure 5.7). This is because the carpal bones in the palm are different lengths. This

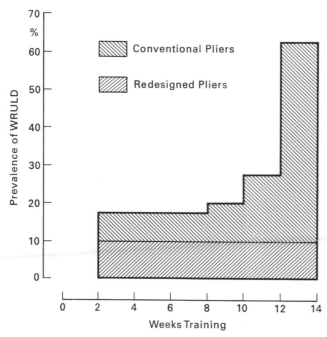

Figure 5.8 Incidence of repetitive strain injuries of the forearm in groups of trainee electronics assembly workers using conventional pliers (upper) and redesigned pliers (lower). Figures for tenosynovitis, carpal tunnel syndrome and epicondylitis (tennis elbow) have been lumped together. Data from Tichauer (1978).

so-called 'natural angle of the wrist' is seen in the traditional designs of carpenters' saws (for example). When using such a tool, the cutting edge of the blade is parallel to the axis of thrust of the forearm when the wrist is in its neutral position. Thus the neutral position of the wrist is preserved in use (see Pheasant 1991a for a further discussion).

Tichauer (1978) describes a study of trainee electronics assembly workers who were required to use pliers extensively for twisting bundles of cables. These trainees had a particularly high incidence of work-related upper limb disorders. Tichauer argued that it was better to bend the handle of the tool than to bend the user's wrist. Figure 5.8 shows the incidence of these conditions in a batch of trainees using conventional pliers, as compared with trainees using the special pliers which Tichauer designed for the purpose. The results are very convincing indeed. This study is now around twenty years old and has been very widely quoted in the literature. To the best of this author's knowledge, Tichauer's redesigned pliers are still not on the market – at least, I have never seen a pair, only pictures of them in books.

CHAPTER SIX

Ergonomics in the office

The basic office workstation typically consists of a desk and chair at which the user will undertake:

 (i) *paper-based tasks*, reading and writing, etc.

(ii) *screen-based tasks*, involving keyboard use (and other input devices).

A few years ago it was confidently predicted that the 'paperless office' was just around the corner – and that before long all information would be handled solely by means of electronic media. Although this goal has not yet been achieved, office work has, over the last decade or so, become steadily more screen-based, to the extent that the office workstation that does not have a visual display unit (VDU) is now a rarity.

It is widely accepted that in the interests of comfort (and the avoidance of adverse long-term effects on health, etc.):

- For writing, the working surface (i.e. desk top) should be somewhat above the user's elbow height, as measured in the standard upright sitting position (see sections 2.4 and 3.8). This is because, in order to write with a relaxed and natural action, the arms must be abducted and flexed somewhat at the shoulders (i.e. raised sideways and forwards).

- For keyboard work, the shoulders should be relaxed with the upper arms hanging freely at the sides and the forearms more or less horizontal and the wrists as far as possible in a neutral position (i.e. neither bent forwards, backwards or sideways). Thus the so-called 'home row' of keys (ASDFG, etc.) should be at or close to elbow height.

- In either case, for comfortable sitting, the user's thighs should be approximately horizontal, with the lower legs vertical and the feet resting flat on the floor. Thus the seat should be at or close to the user's popliteal height – but preferably somewhat below it.

- In the case of keyboard work there is an element of controversy as to whether it is preferable that the user's trunk should be upright or reclined. We shall return to this last point in due course (see section 6.5 below).

In the UK, office furniture and workstations are the subject of two British Standards: BS 5940 (1980) which deals with furniture, and BS 7179 (1990) which deals with visual display terminals. At the time of writing, the latter is in the process of being superseded by a European Standard BS EN 29241 (1993) which is in turn based upon the International Standard ISO 9241 (1992). In the UK, such standards are regarded as advisory but not mandatory. Screen-based office work, however, is subject to the provisions of the 1992 *Health and Safety* (*Display Screen Equipment Regulations*) which implement an EC Directive on the subject. The reader who is concerned with compliance with these regulations is referred to the relevant HSE publication (HSE 1992a), in which they are discussed in detail.

6.1   The office desk

The user has three points of physical contact with his workstation and environment: the desk (or keyboard), the seat and the floor. If a range of users, who vary in both size and shape, are to attain the same desirable working position (as defined by the anthropometric criteria which are set out above) then two out of these three must be adjustable. Nowadays, virtually all office chairs are adjustable for height. In the UK, however, adjustable office desks remain something of a rarity (although there are some signs of gradual change in this respect at the time of writing). Elsewhere in the world adjustable desks are more common. This is the case for example, in Australia, where the so-called 'RSI epidemic' of the 1980s (see Chapter 8) acted as a major stimulus to the improvement of working conditions in offices.

Ergonomically, adjustable height desks are the preferred solution for office work, particularly if this is intensively screen-based. A fixed height desk may be regarded as an acceptable second best, provided that the floor level is adjustable, which in practice may be achieved (in part) by the provision of footrests where required.

Given that the desk is to be of fixed height, what height would represent the best possible compromise for a working population of adult men and women? Supposing we say that for *paper-based* office work the desk top should ideally be 75 mm above the user's sitting elbow height, and the seat should be 50 mm lower than his or her popliteal height (PH). Then optimal desk height = (SEH + 75) + (PH − 50). Since we are concerned with finding the best possible compromise (which minimizes the number of people falling outside whatever tolerance bands about this optimum we care to propose) then 50th %ile anthropometric values are required (see section 2.3). On this basis the best single compromise desk height for men would be 735 mm and for women 705 mm – giving an overall figure of 720 mm.

The British Standard for Office Furniture (BS 5940, which was published as long ago as 1980) specifies a height of 720 ± 10 mm for a 'general purpose office desk'. Given that the office work of the day was almost entirely paper-based, by the above calculations this would seem to be about right.

For *keyboard work*, however, the anthropometric criteria given above indicate that the best possible compromise height would be a good deal lower: first, because the home row of the keyboard should be at sitting elbow height rather than above it; and second, because the home row will itself be some 30–50 mm above the desk surface. But the difference may not be as important as it seems, in that although we may in theory regard a seat height of 50 mm below popliteal as ideal, in practice, a seat of popliteal height or even a little above would probably be entirely acceptable to the majority of users. In reality therefore, a standard desk height of 720 mm or thereabouts is probably just about as good a compromise as any other for screen-based office work – although we must bear in mind that, as a compromise, it is somewhat biased in favour of the taller half of the user population. This, however, as we shall see, is probably no bad thing.

(A figure of 720 mm is given in BS 7179 as the 'recommended' height for the surface on which the keyboard is placed. The part of ISO 9241 which deals with these matters is not yet available at the time of writing.)

Office desks are made to a standard height; office workers are not. The standard desk is satisfactory for the *average person* (see Table 1.1). But for people who are markedly shorter or taller than average it can cause serious problems (particularly for intensive screen-based keyboard work, where the potential long-term consequences of a mismatch are likely to be greater).

To reach an appropriate working height in relation to the keyboard, a girl with short legs working at a standard height desk will have to adjust her chair to a level that is too high for comfort. As a consequence she will tend to perch on the front edge of the seat, thus losing the support of the backrest. This may lead to back problems and worse (see Section 6.4 below). If she lowers the seat, however, she will commonly end up working with her shoulders hunched and her arms abducted (i.e. the elbows raised out sideways). The static muscle loading which results may lead to neck and shoulder problems. Abduction of the arms at the shoulders calls for a compensatory ulnar deviation of the wrists, in order to maintain the alignment of the fingers at the keyboard (i.e. the wrists are bent sideways in the direction of the little finger). This is very unsatisfactory indeed, the ulnar deviated wrist being an important causative factor in the aetiology of upper limb disorders (see Chapter 8). Alternatively, instead of abducting her arms, the short user may incline her forearms upwards, which leads to flexion (i.e. forward bending) of the wrists, or she may work with her wrists in extension (i.e. backwards bending), both of which are undesirable.

The problems of the short user are solved relatively easily with a footrest. As a rough guide, anyone of around 5′3″ (1600 mm) or less in stature will probably need a footrest when working at a standard height desk.

The problems of the unusually tall user are more difficult to solve. At a standard height desk he will tend to find himself working with his spine in flexion however he adjusts his chair. The only real solution is a non-standard desk or one of adjustable height – or a standard desk which is raised up in some way.

In the final analysis, an adjustable desk may be regarded as the ergonomically preferable solution to the postural problems of the keyboard user. The range of adjustment required may be calculated from the combined dimension shod popliteal height + sitting elbow height; which has a distribution of 710 [42] mm in men and 690 [40] mm in women. We may calculate therefore, that for keyboard work the desk top should adjust from about 600 mm to 750 mm to accommodate the 5%ile female and 95%ile male user (assuming a keyboard thickness of 30 mm); and for writing the range should extend somewhat higher. In practice, however, the bottom part of this range may be of limited use because of knee-room problems. BS 5940 specifies a range of 670–770 mm and BS 7179 specifies 660–770 mm. In practice this is probably about right.

The distance between the undersides of the elbows and the tops of the thighs in the 'standard sitting position' (see Section 2.4) is only around 80–85 mm on average – and in many people it will be a good deal less. If the user is to adopt the recommended keying position, as described above, this space must accommodate the thickness of the keyboard (generally some 30–50 mm on a modern machine) plus that of the desk top itself. It follows therefore that the recommended keying position will be a physical impossibility for a proportion of users. It is a matter of utmost importance furthermore that the thickness of the desk top (and its supporting structures) should be kept down to an absolute minimum, commensurate with the requirements of structural strength. In particular, desks that have obstructions below the working surface, such as 'kneehole' drawers, etc., must be regarded as wholly unsuitable for keyboard work. (See the story of Janice at the beginning of Chapter 1.)

An important but frequently neglected ergonomic aspect of desk design is the adequacy of its surface area. The desk top must be large to allow the screen to be placed at a suitable viewing distance (see below) and permit the user some degree of flexibility in where she places the keyboard. The overall space that is needed will of course depend also on what other items live on the desk. Clutter expands to fill the available space (Pheasant's principle of ergonomic decay).

Dimensional recommendations for office desks are summarized in Figure 6.1.

6.2   The office chair

6.2.1   Seat height

To meet the requirements of a range of users the height of the seat should be easily adjustable from the sitting position. The height range that is required will in principle depend upon whether the seat is to be used with an adjustable desk or a fixed height desk (presumably of standard height). If the former, then a range of 5th %ile female to 95th %ile male, shod popliteal height, would seem appropriate – which works out at 380–515 mm (assuming 25 mm heels for both sexes). Given a 720 mm desk, it is unlikely that anyone will want a seat higher than 535 mm (720 minus 5th %ile female sitting elbow height). A height range of 380–535 mm should thus in principle meet all eventualities. In practice it may be a little over-generous.

6.2.2   The backrest

Typists' chairs traditionally had low-level backrests, whereas executive chairs had medium-level or even high-level backrests. The supposed justification for this was that a typist needed freedom of movement for her shoulders. In reality, however, it was probably more a matter of the differentiation of status – combined perhaps with a puritanical distrust of comfort in the workplace. With the old-fashioned mechanical typewriter, the argument for the low-level backrest was perhaps just plausible; with the modern electronic keyboard it is no longer valid. A medium-level backrest gives better back support and permits a more reclined (and thus more relaxed) working position (see below). Grandjean (1987) recommends a 500 mm backrest.

In order to give the user the greatest possible variety of working positions, the angle of the backrest should be adjustable (independently of the seat). The backrests of many modern office chairs are spring loaded such that they follow the user's changes of position. In theory this seems like a good idea. In practice, some users like it; some do not. So it is important that the user should also be able to lock the backrest in place if he or she wishes.

Finally, the backrest should be contoured to the form of the lumbar spine and adjustable in height (again relative to the seat) so that the user may match the mid-point of the seat's lumbar pad to the curve of his or her own back. As we noted earlier it is important that the 'contour' of the backrest should not be excessive. Some modern office chairs are definitely 'over the top' in this respect (see also the discussion of backrests in Section 4.3).

A brief comment is called for at this point, concerning the HSE 1992 Display Screen Equipment Regulations (HSE 1992a). The original EC Directive upon which these regulations are based states: 'The seat shall be adjustable in height. The seat back shall be adjustable in both height and tilt'. The same words are repeated in the 'Schedule' of the HSE Regulations which deals with minimum requirements for workstations. What would a reasonable person suppose this means? Surely it must mean that the seat back should be adjustable relative to the seat surface – otherwise there would seem little or no point in commenting on the matter at all. The HSE Guidance Document (HSE 1992a) interprets it as meaning that the backrest should be adjustable for height and tilt relative to the ground, and says there is no necessity for it to be adjustable relative to the seat. In ergonomic terms, this interpretation is (in the opinion of this author) greatly regrettable.

6.2.3 Seat tilt

Some modern office chairs incorporate a rocking mechanism in the seat such that it may be tilted forwards and backwards. Bendix and Biering-Sorensen (1983) report a trial in which it was found that subjects preferred a seat that was free to tilt between an angle of 5° forwards and 5° backwards compared with seats fixed in either position. Experience indicates, however, that many users actively dislike tilting seats; so again it is important that the user should be able to lock the tilt mechanism in place if he or she wishes. In the author's personal view there is little to be said in favour of tilting seats – but not everyone would agree with me.

6.2.4 Armrests

Traditionally, typists' chairs did not have arms whereas 'executive' chairs commonly did. As with backrests (see above) this was in part a matter of ergonomics and in part the differentiation of status. Some keyboard users like to support their elbows on the arms of a chair as they work – and insomuch as it reduces the static loading on the muscles of the neck and shoulder girdle this would seem no bad thing. An alternative which achieves the same end is to support the wrists (see Section 6.6 below). Armrests can be a mixed blessing, however, if they prevent the user from getting close up to the desk.

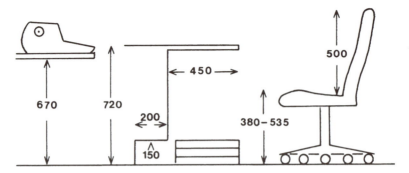

Figure 6.1   Compromise dimensions for office furniture (in mm).

6.2.5   The usability of adjustment controls

The more different ways it is possible to adjust a chair, the more difficult it becomes to design adjustment mechanisms that are easy to operate. The more difficult an adjustment mechanism is to operate, the less likely it is that it will be adjusted properly. As a general principle each mode of adjustment should have its own dedicated control lever. Coupled adjustment mechanisms, in which the backrest angle and seat tilt are controlled by the same action, are especially undesirable.

Office workers frequently do not know how to operate the adjustment mechanisms of their chairs. Teaching them how to do so may often improve their comfort dramatically.

Dimensional recommendations for office chairs are summarized in Figure 6.1.

6.3   Visual demands of screen-based work

For visual comfort in screen-based work a viewing distance of 500 mm may be regarded as an absolute minimum and 750 mm would probably be better (see Section 3.7). The material displayed on the screen should thus be designed in such a way as to be legible at an adequate distance. This is partly a matter of character size and partly one of image quality. A modern high-resolution screen has a degree of legibility that approaches that of printed text. But older screens may be very much less legible (particularly if the image is unstable due to technical faults, etc.). There is evidence that dark characters on a light ground are superior in this respect to light characters on a dark ground (Radl 1980, Bauer and Cavonius 1980). Both the brightness and the contrast of the screen should be separately and easily adjustable by the user. Lighting is also a factor. In general glare from light sources will be minimized if the screen is placed at right angles to a window, and parallel to overhead fluorescent tubes. Diffused uplighting is better, however. (For a more detailed discussion see, e.g., Pheasant 1991a,b.)

It is generally agreed that the screen should be a little below eye level, such that when looking at its centre the user has a downward visual angle of about 15° (see also Section 3.7). This will eliminate the poked chin and rounded back of the 'yuppie hump' (Figure 6.2) which comes from looking at a screen which is too low. (The legibility of the screen may also be a factor in causing the poked chin.) When

Figure 6.2 VDU user demonstrating the 'yuppie hump', from an original in the author's collection. (From S. Pheasant, *Ergonomics, Work and Health*, Macmillan, 1991, fig. 5.4, p. 111, reproduced with kind permission.)

looking at the centre of the screen the user's line of sight should be approximately perpendicular to the screen surface. To meet these requirements, it will be necessary for the screen to be physically separate from the keyboard; to tilt and rotate on its base; and to be designed in such a way that it can if necessary be raised on a plinth or some other suitable support.

6.3.1 Document holders

The provision of a holder for source documents (so that they can be read at screen level and without turning the head) will reduce the postural loading on the neck muscles very considerably (particularly in data entry, copy typing tasks, etc.). The preferred location of the document holder, relative to the screen, depends upon the task. In general, the item that the user looks at more frequently should be placed directly in front of him. If screen and source documents are referred to equally often, then they should be placed on either side of the straight-ahead position and angled slightly inwards towards each other.

6.3.2 The unskilled keyboard user

The two-finger 'hunt and peck' typist presents a special set of problems, since he will inevitably spend much of his time looking at the keyboard itself, rather than the screen or source documents. He will tend to work in a hunched position. The problem can be ameliorated to some extent by encouraging him to adopt a reclined position and be as relaxed as possible. Beyond this, it is difficult to see what else can

Figure 6.3   The laptop computer. Note the posture of the neck.

be done – other than teaching him touch typing (which for anyone using the keyboard on a habitual basis is clearly desirable).

6.3.3   The laptop

The increased use of laptop computers to save desk space is currently becoming a cause for concern in some circles. The problem is, first, the height of the screen, and second, that it is only legible when viewed through a narrow range of angles. Together these factors fix the head and neck in an unsatisfactory position. The problem may not be quite as bad as it seems, insomuch as the screen of a laptop may be tilted back to any angle the user chooses. The typical postural adaptation that occurs therefore is a simple inclination of the head (Figure 6.3). Although this is by no means desirable (see Section 3.7 above), it is probably not as bad as the yuppie hump (which comes from looking at a *vertical* screen which is too low).

6.4   The keyboard (and other input devices)

As we noted above, the keyboard should be as thin as possible. It is usually recommended that its slope should be adjustable, although in biomechanical terms a very slight rake would seem preferable to a steeper one.

Users often have strong views about the 'feel' of different keyboards. One user once described the difference between working at two particular makes of keyboard as 'like the difference between walking on turf and walking on pavement'. At present, however, we know little or nothing about these matters on a formal scientific basis. The features of the keying action, which give the keyboard a 'good feel', remain elusive. Different users seem to have different (and sometimes conflicting) views on the subject.

There is good evidence that neither the conventional QWERTY keyboard layout, nor the presentation of the alphabetical keys in a single rectangular key field on a plane surface, is an optimal ergonomic solution to the problem of keyboard design (see Pheasant 1991a,b, for a detailed discussion). The former issue is to all intents and purposes a lost cause. There may be more realistic opportunities for improvement in terms of the latter. The basic problem with the plane keyboard and rectangular key field is that a degree of ulnar deviation will inevitably be required to maintain the alignment of the fingers at the keys. (This is accentuated if the keyboard is too high, see above; and if the user leans forward, see below). A number of 'split keyboard' designs have been proposed to overcome this problem. Initial trials of these seem encouraging (see, e.g., Grandjean 1987), but as yet they show little or no sign of catching on.

6.4.1 Other input devices

We may confidently predict that as time goes by the keyboard as such will increasingly be replaced by other devices for entering information onto the machine and otherwise controlling its functions. The extent to which these will supplant the keyboard in everyday applications – or perhaps more realistically the rate at which they will do so – is impossible to estimate at present.

The most widely used input device other than the keyboard at the present time is the 'mouse'. There are already signs that people who use the mouse extensively (in layout work, etc.) are prone to suffer from upper limb disorders similar to those of keyboard users – the causative factors presumably being the particular combinations of static muscle loading due to working posture and the repetitive motions of the wrist and/or fingers that the tasks in question entail.

The ultimate alternative to the keyboard (short of psychokinesis) is voice input. It will doubtless be found to cause RSI of the vocal chords.

6.5 'Good posture' in screen-based work

Concern over the increased reporting of work-related upper limb disorders in the white-collar population (see Section 8.5) has led to an increasing demand that keyboard users should be taught 'the correct way to sit'. To the layperson this requirement may seem straightforward enough. It is not quite as simple as it seems, however, and a number of unresolved issues remain. There are two basic schools of thought concerning the matter. We could call these the orthodox or 'perpendicular' approach, and the alternative or 'laid-back' approach. The postures in question are illustrated in Figure 6.4. The views that people take on these matters are not always stated entirely explicitly – and indeed the whole issue is surrounded by something of an air of vagueness. Overall, the adherents of both schools would be in a broad measure of agreement concerning the anthropometric criteria proposed at the beginning of this chapter. They would also agree that working postures that entail

Figure 6.4   Working posture at the visual display terminal as recommended by Cakir *et al.* (1980) (left) and Grandjean *et al.* (1987, 1984) (right).

forward leaning (and in particular the yuppie hump) are highly undesirable. The principal points at issue are, first, the desirable position of the trunk, and second, whether or not the wrists should be supported.

The *perpendicular position* for keyboard work (in which the trunk is as far as possible kept upright with the back principally supported in the lumbar region so as to maintain its 'normal' curve) has, to this author's knowledge, been taught in schools of typing since the 1930s at least. It may be regarded as representing the 'conventional wisdom' on the subject, which until the mid-1980s or thereabouts would have commanded more or less universal consent. People who still teach the perpendicular position sometimes refer to it as 'sitting in balance'. This sounds very good. But it does not really get us much further forward in understanding the issues involved, in the absence of a formal explanation of what is meant by 'balance', framed in the language of physiology and biomechanics. Such an explanation, in this author's experience, has not been forthcoming.

The first significant challenge to the orthodox view was made by the late Professor Etienne Grandjean and his co-workers in Zurich, in a series of papers published in the early 1980s, the findings of which are summarized in his book *The Ergonomics of Computerized Offices* (1987). Grandjean's views were based upon trials in which VDU users were provided with fully adjustable chairs and workstations to use in their own offices. The great majority of subjects preferred a 'laid-back' position in which the trunk was reclined by between 10° and 20° to the vertical. (Only about 10% of subjects chose to sit upright.) Where a padded wrist support was available, the great majority of subjects (80%) chose to use it; and where it was not available around half of the subjects chose to rest their wrists on the desk. On average the subject's elbows were flexed to a little less than a right angle, so the forearms were inclined slightly upwards.

Grandjean (1987) argues forcefully that in biomechanical terms there is nothing whatsoever wrong with this position. He mainly bases his views on the experimental evidence of Anderson *et al.* (1974, cited in section 4.2 above) who showed that when

Figure 6.5   Keyboard worker in a natural, relaxed position.

the trunk is so reclined, the loading on the lumbar spine is substantially less than it is when sitting upright. He does not, however, explicitly address the issue of upper limb disorders in this context.

On balance, a supported wrist when using the keyboard would seem to be desirable rather than otherwise, in that it will reduce the static loading on the muscles of the neck, shoulder and arm. There are two caveats to this general position. One is that supporting the wrist on the sharp edge of the desk (which you see quite commonly) can cause blunt trauma to the tissues of the front of the wrist (and in particular the ulnar nerve). The second is that it may result in a 'cocked' (i.e. extended) wrist, which results in a static loading of the muscles in the extensor compartment (i.e. back) of the forearm. The latter is particularly likely to be a problem if the keyboard is abnormally thick or if it is used in a steeply raked position. Both are highly undesirable. Both may be avoided by the use of a padded wrist support.

In addition to reducing the mechanical loading on the lumbar spine, the laid-back sitting position (particularly when combined with the supported wrist) has the additional advantage of tending to increase the overall horizontal distance between the user's shoulders and the keyboard. (The disadvantage that would accrue in terms of static load is eliminated by the supported wrist.) It follows, as a matter of geometry, that the degree of wrist deviation required to maintain the alignment of the fingers on the keys will be correspondingly diminished. As the trunk moves from a reclined position to an upright position, and then from an upright position to a forward sitting position, the elbows must also move out sideways to accommodate the width of the lower part of the rib cage. This results in a further and progressively more pronounced ulnar deviation of the wrist.

Opponents of the laid-back approach argue that it tends to degenerate into a slumped position similar to that of the yuppie hump. By the same token, the upright position will tend to degenerate into a forward slump. With a well-designed seat that gives good back support, the former tendency will in my view be minimal; whereas the latter will still be present to a more marked extent, particularly when the user is tired or under stress.

On the basis of these various considerations therefore (and notwithstanding my comments on the subject in earlier writings), I am currently of the opinion that the laid-back position for keyboard work offers material advantages as compared with the perpendicular position. Having said this, I remain disinclined to be unduly prescriptive in advising the individual keyboard user as to what constitutes a 'good posture'. It is more important that he or she should learn the importance of postural diversity in the workplace and the avoidance of unnecessary muscle tension. The laid-back position is only desirable in that it is materially more likely to achieve these aims.

Figure 6.5 (which was taken from an unposed original) shows a keyboard worker in a natural relaxed sitting position at a well-designed fully adjustable workstation.

6.6   The design of screen-based working tasks

It is widely recognized that prolonged periods of intensive screen-based keyboard work (particularly repetitive tasks such as data entry and copy-typing), unbroken by rest pauses or changes of working activity, are of themselves highly undesirable, and have the potential to result in musculoskeletal injury. The greater the 'exposure' to intensive keyboard use, the greater the risk, however good the ergonomics the workstation and working posture; the worse the ergonomics, the greater the level of risk for a given level of exposure (see also Chapter 8).

In organizing office work therefore (and in allocating the various tasks to be performed in the office to individual workers), periods of intensive screen-based keyboard use should, wherever possible, alternate on a frequent and repeated basis with other working tasks of a contrasting nature. Where this cannot reasonably be achieved (as regrettably is all too often the case in the electronic sweat shops of the data entry trade) a suitable daily schedule of rest pauses should be set up. But this is by way of being second best.

It is difficult to be precise about these matters. As an approximate rule of thumb, 'screen breaks' of around 5 min in each continuous period of keyboard work would seem about right – these being taken *in addition to* the person's normal lunch break and mid-morning and mid-afternoon coffee breaks. The office worker should be actively encouraged to get up, move around and stretch during these breaks – and the habit of taking lunch at the desk is strongly to be deprecated. 'Micropauses', in which the user consciously stops work to relax, for a few seconds every few minutes, are also greatly to be encouraged.

In other words, if an adequate degree of task diversity cannot reasonably be achieved, then the working period should be structured in such a way that it is broken up with micropauses, short pauses, and long pauses, in which recovery from fatigue can occur.

CHAPTER SEVEN

Ergonomics in the home

The house and home may in general be divided into a number of more or less discrete spaces, each of which is specialized for the performance of particular range of purposeful activities, which in a broad sense we could call 'working tasks'. Most typically the boundaries of these spaces will coincide with the rooms of the house. The spaces may overlap physically to some extent, and tasks may intrude to some extent from one space to another. In this chapter we shall consider the ergonomics of three such 'workspaces': the kitchen, the bathroom and the bedroom.

7.1 The kitchen

Of the specialized spaces of the home, the kitchen is the one that can most obviously be treated as a functional working area, in the ordinary narrow meaning of the word 'work' – and it is perhaps because of this that it is the one that has been discussed most extensively in the ergonomics literature. We shall consider, first, the layout of the floor plan, and second, the heights of working surfaces and other related matters.

7.1.1 Layout

The literature on this subject reveals two basic design principles, which turn out to be variants of McCormick's sequence-of-use and frequency-of-use principles, respectively (see Table 3.1).

(i) For a right-handed person the sequence of activity proceeds from left to right thus: sink to main worksurface to cooker (or hob) to accessory worksurface for 'putting things down'. It clearly makes sense for this sequence to be unbroken by tall cupboards, doors or passageways; but it need not be in a straight line – an L- or U-shaped configuration will serve just as well. Another accessory worksurface left of the sink completes the layout (Figure 7.1).

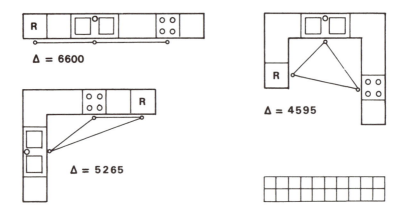

Figure 7.1   Three kitchen layouts designed, according to the principles discussed in the text, on a 300 mm modular grid. Each includes refrigerator (R), cooker and similar lengths of working surface. The work triangle is indicated and its length (Δ) given.

(ii) The refrigerator (or other food store such as larder, freezer, etc.), sink and cooker, constitute the much discussed 'work triangle' of frequently used elements. For reasons of safety, through circulation should not intersect this triangle – particularly the route from sink to cooker which is used more than any other in the kitchen. The sum of the lengths of the sides of the triangle (drawn between the centre front of the appliances) should fall within certain prescribed limits. Grandjean (1973) quotes a maxima of 7000 mm for 'small to medium-sized kitchens' or 8000 mm for 'large kitchens'. Department of the Environment (1972) gives a minimum combined length of 3600 mm and a maximum of 6600 mm for the kitchen 'to leave adequate working space and yet be reasonably compact' and also specifies that sink–cooker distance should be between 1200 and 1800 mm in length (Figure 7.1).

7.1.2 Worktop height

In order to determine an optimal height for kitchen working surfaces, we must consider both the anthropometric diversity of the users and the diversity of tasks to be performed. If, as is commonly the case, a 175 mm deep sink is set into the worksurface, then the effective working level may range from perhaps 100 mm below the worktop height when washing up to a similar distance above it when operating machinery or mixing with a long-handled spoon. We should expect differences, even among tasks performed upon the surface itself, associated with varying requirements of downward force, e.g. between rolling pastry and spreading butter.

Ward and Kirk (1970) studied these matters by means of a fitting trial. The subjects, who were all women, performed three groups of tasks and selected the following worktop heights as optimal:

- group A – tasks performed above the worktop (peeling vegetables, beating and whipping in a bowl, slicing bread), 119 [47] mm below elbow height;

- group B – tasks performed on the surface (spreading butter, chopping ingredients), 88 [42] mm below elbow height;
- group C – tasks involving downward pressure (rolling pastry, ironing), 122 [49] mm below elbow height.

These results were subsequently confirmed using a variety of physiological measurements (Ward 1971) in which it was also shown that the optimal height for the top edge of the sink was approximately 25 mm below the elbow.

The next stage in the analysis is to allocate priorities to these three groups of tasks. Ideally, this would best be done by means of an observational survey of user behaviour. By way of second best we ask a sample of 'typical' kitchen users – and find a general agreement that the group B tasks are more important and the group C tasks least important. Allocating a weighting of 4 to group B and a weighting of 1 to each of the others we arrive at an overall recommendation of 100 mm below elbow height for the optimum height of the worktop. Combining this with anthropometric data for our standard population (see Section 2.4) gives us the figures set out in Table 7.1.

The dimensional co-ordination of kitchen equipment is obviously desirable, and the purchaser should have every confidence that a new oven will fit into the existing range of units. The provision of adjustable worktops, or fittings in a range of heights, is not incompatible with this goal – but it does obviously make things more difficult (and therefore costly). In response (presumably) to Ward's studies, BS 3705 (published in 1972) reads thus: 'Subject to the need for field research and solving technical problems, it is thought that a 50 mm incremental range of heights of working surfaces may be adopted in the future, the ranges being 900 to 1050 mm for sinks and 850 to 1050 mm for worktops. Because studies show that generally the worktop surface needs to be higher than the present 850 mm for the greater number of users, this standard is omitting the 850 mm worktop height, although this might be included in any subsequent range after the above research is completed. As an interim measure the standard will remain at the BS 3705 (imperial) height, rounded off metrically to 900 mm for sinks and worktops.

A decade later the 'interim measure' had acquired a distinct air of permanence. BS 6222 (published in 1982) reads: 'The co-ordinating heights of all units and appliances shall be as follows ... top of worktop: either 900 or 850 mm (second preference)'. ISO 3055 (published in 1985) also specifies a standard worktop height of 850 or 900 mm – but hedges its bets in an annex (which is not regarded as part of the standard proper) by referring to 'appropriate' heights of 850 to 1000 mm for

Table 7.1 Optimum heights of kitchen sinks and worktops.

	Men			Women		
	5th %ile	50th %ile	95th %ile	5th %ile	50th %ile	95th %ile
Worktop	930	1015	1105	855	930	1005
Sink	1005	1090	1180	930	1005	1080

food preparation and 900 to 1050 mm for washing up, suggesting that: 'Adjustments can be different plinth heights and other means'.

Figure 7.2 shows a set of standard kitchen units in side elevation as compared with the height optima from Table 7.1. The 900 mm worktop is lower than ideal for approximately half of women and virtually all men. The sink (with its rim at the same height as the worktop) is too low for just about everyone.

Compare this state of affairs with that of the standard office desk, which we discussed in the last chapter (section 6.1). Our analyses indicate that the 900 mm standard is very far from being the best single compromise height (and the 850 mm, which fortunately does not seem to be used in the practice, would be that much worse).

How serious a problem is this in practice? This question (as is always the case with such matters) is the more difficult to answer. The potentially deleterious effects on the tall user, of working at a sink or worktop that is too low, will depend to a considerable extent on how long he or she does it for at a stretch. (Compare this again with the matter of 'exposure' to intensive keyboard use discussed in Section 6.6.) For those of us who are fortunate enough to have sound backs, a sink or worktop that is too low will be no more than a minor annoyance – which will in all probability go more or less unnoticed. In other words, it is something that we are readily able to adapt to. (The third fundamental fallacy – see Table 1.1.) For those of us who are less fortunate in this respect – and low back trouble is very common indeed – the task of washing up or preparing a meal at a sink or worktop that is too low may be an intensely painful experience. At the point at which it becomes intolerably painful the person is effectively 'disabled' in respect of this task.

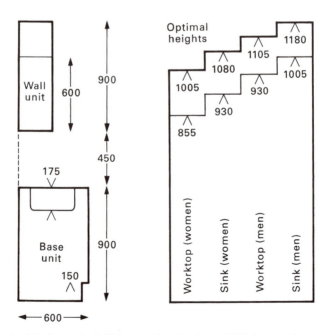

Figure 7.2 Standard kitchen unit: (left) base and wall units and (right) optimal ranges of worktop heights (5th–95th %ile). Analysis of storage space according to the criteria of Table 7.2.

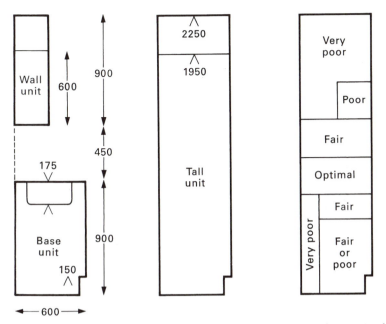

Figure 7.3 Standard kitchen tall unit. Analysis of storage space according to the criteria of Table 7.2.

7.1.3 Storage

It is instructive to compare side elevation views of standard kitchen units with ergonomic recommendations for storage zones, (as discussed in the next chapter, see Table 8.4) – as shown in Figure 7.3. Cupboard space in the 800–1100 mm optimum height range being strictly limited, the most accessible storage space in the kitchen becomes the worktop – which disappears under a clutter of food mixers, spaghetti jars and other homeless objects. Clutter expands to fill the available space – Pheasant's principle of ergonomic decay (see also Section 6.1).

7.2 The bathroom

The bathroom should combine hedonistic luxury with functional efficiency. It is an environment in which to relax and unwind, soaking in a hot tub, but also a configuration of workstations for the practical activity of washing, grooming and excretion (assuming a special room is not set aside for the latter). *The Bathroom* by Alexander Kira (1976) is a classic of user-centred design research, which every interested person should endeavour to read.

7.2.1 The bathtub

The bathtub presents interesting problems of dimensional optimization. It must be large enough for comfortable use by one person (or perhaps two) but should not have needless volume, requiring filling with expensively heated water. It is also a notoriously hazardous environment for the frail and infirm.

Two principal postures are adopted in the bath: a reclined sitting position and a recumbent position (possibly with the knees flexed) in which the body is submerged to neck level. For comfort in the sitting position the horizontal bottom of the tub must be sufficient to accommodate buttock–heel length (95th %ile man = 1160 mm) and the end of the bath should provide a suitable backrest. Kira (1976) recommends a rake of 50–65° from the vertical and contouring to conform to the shape of the back. This seems excessive to me – a rake of 30° and a suitable radius where the base meets the end should be quite adequate – we are not particularly looking for postural support since the buoyancy of the water will both unload the spine and lift it away from the backrest. The more we increase the length of the horizontal base the greater the possibility for total submersion. We may shorten our recumbent bodies by around 100 mm by flexing our knees – given that we wish to keep our heads above water – and that as the 95th %ile male shoulder height is 1535 mm there seems little point in lengthening the horizontal part beyond around 1400 mm.

The width of the bath must at least accommodate the maximum body breadth of a single bather (95th %ile man = 580 mm). The ergonomist, who is usually a broad-minded sort of person, should also consider the accommodation of couples. Methods for calculating the combined dimensions of more than one person are discussed in section 4.3. For couples wishing to sit side-by-side the necessary clearance is given by their combined shoulder breadth (920 mm for a 95th %ile couple of the opposite sex). For couples sitting at opposite ends (probably the more common arrangement) the clearance is given by combined breadth of the hips of one person and the feet of the other. This is greatest when the hips are female and the feet are male, in which case the 95th %ile combination is 625 mm. This arrangement does, however, demand that the taps should be in the centre to avoid arguments (Figure 7.4).

Consider a 95th %ile man (sitting shoulder height 645 mm) reclining against the end of the bath. His shoulders will be 645 cos 30 = 558 mm above the base of the bath. He could not reasonably require more than 400 mm of water. If the backrest was raked further, say to 45°, then 300 mm of water would suffice. (These figures are

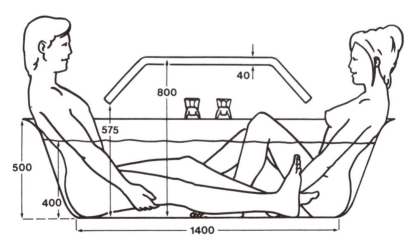

Figure 7.4 The ergonomically designed bath.

mere speculation; it would be very interesting to perform a fitting trial to find out what depth of water people really do want.) Assuming a 30° rake, a bath depth of 500 mm would be required for an adequate quantity of water without too much danger of it splashing over the edge. In fact, a typical tub depth at the present is about 380 mm (15 in.), although older models are often deeper. The outside height of the rim (above the floor) is, of course, generally greater than the tub depth (often by as much as 100 mm).

A deep bath or a high rim is generally deemed to make entering and leaving more difficult and hazardous – although Kira (1976) casts some doubt on this, arguing that the manoeuvres people use to enter and leave baths have been insufficiently analysed. Grandjean (1973) cites evidence that a height of 500 mm is acceptable to most elderly or frail people. Grab rails are usually advocated as an aid to stability. These could reasonably be a little above knuckle height at the point where you climb in (e.g. 800 mm above the bath base), around shoulder height (e.g. 575 mm) at the sitting end and about 40 mm in diameter. Vertical grab rails may well be better for the infirm. Additional holds along the side of the bath are also desirable and, for the frail, a non-slip mat inside the bath is essential.

7.2.2 The handbasin

This device will be used for washing the hands and face and sometimes the hair. The criteria are relatively simple: it should be possible to wet the hands without water running down the forearms and bending should be minimized. Hence, a basin rim that is at about the elbow height of a short user would be appropriate (5th %ile woman = 930 mm unshod). Kira (1976) studied the above activities experimentally by observing subjects first miming the actions without the constraints of an appliance and then using an adjustable rig. On the basis of these fitting trials he concluded that, for washing the hands, the water source should be located some 100 mm above the rim of the basin, which should be set at 915–965 mm. Conventional handbasins are very much too low (commonly less than 800 mm) – except perhaps for use by children. The present practice of placing the taps at or below the level of the rim seems based on the assumption that people will fill the bowl and wash in the water therein. In fact, according to Kira, 94% of people prefer to wash under a running stream of water.

7.2.3 The water closet

There is a strong body of opinion that takes the view that the sitting posture, which in Western society we use when emptying our bowels, is physiologically unsound. Proponents of this view – most notably Hornibrook (1934) – argue that a squatting position, in which the thighs are pressed against the abdominal wall, encourages an easy and physiologically more efficient bowel movement, which in the long run will help prevent a variety of nasty diseases (to which we are prone as a result of our diet and sedentary habits).

I am not aware of this theory having been tested experimentally. (It would be difficult to do so.) But experience suggests that it is basically correct. The physiology remains unclear. It is not solely a matter of increased intra-abdominal pressure, as has been suggested – since this is the act of 'straining at stool' which is allegedly deleterious.

The plinth of the conventional WC is typically around 380–400 mm in height. For squatting this would have to be halved. One consequence of reducing the height is that the buttocks take a much greater proportion of body weight. So the contouring of the lavatory seat itself becomes much more critical for comfort. But as you would not have to sit there for so long it probably would not matter so much.

For a discussion of the ergonomics of the more conventional sorts of lavatory, see McClelland and Ward (1976, 1982).

7.3 The bedroom

Considering the amount of time we spend in bed and the importance of sound sleep to our overall well-being it seems remarkable how little formal scientific study has been devoted to the ergonomics of bed design.

Tall people commonly complain about beds being too short. Noble (1982) cites the results of a survey of beds on the market in the UK. Both single and double beds ranged in length from 1900 mm to 2360 mm. The length of the recumbent body is somewhat greater than stature; and the bed should be somewhat longer still, since people sometimes like to sleep with their hands beyond their heads. Assuming that a person will require a bed length of at least 150 mm greater than his or her stature for comfort, we may calculate that a bed length of:

1980 mm will be too short for 1 man in 10;
2055 mm will be too short for 1 man in 100;
2105 mm will be too short for 1 man in 1000;
2150 mm will be too short for 1 man in 10 000;

and so on (see Section 2.1).

Bed width is more complicated, being not solely a matter of anthropometrics. A sound sleeper may make up to 60 gross changes in posture during the course of a night. Physiologically, these are the equivalent of 'fidgets'. They serve to preserve sleep, by relieving muscle tension, preventing the build-up of pressure hot-spots, and so on – these being potential sources of neural signals of discomfort such as might wake us (see Section 3.6). The bed should be wide enough to allow these changes in posture to proceed unimpeded. In practice this tends to mean the wider the better. There must logically be a point beyond which further increases in width would carry no further benefit. The determination of this point would have to be the subject of empirical studies.

In Nelson's day (c. 1800) the poles of a sailor's hammock were a standard 18 in. (450 mm) long. This, being less than a 95th %ile male shoulder breadth (510 mm), would have permitted many men side-lying postures only. A modern navy bunk is at least half as wide again at 27 in. (685 mm). By way of comparison a typical beach mat is 24 in. (610 mm) wide; and the standard NHS hospital bed is 910 mm. The single beds in the survey cited by Noble (1982) ranged from 750 to 1000 mm in width ($n = 21$), and the double beds from 1200 to 2000 mm ($n = 38$) – see Figure 7.5.

According to Brewer's Phrase and Fable, the legendary *Great Bed of Ware* (*Twelfth Night*, iii, 2) which was said to have belonged to Warwick the Kingmaker, was 12 feet square and capable of holding 12 people. The historical Great Bed of

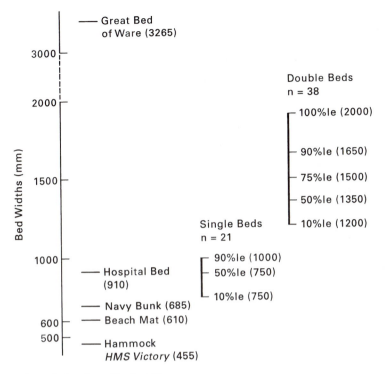

Figure 7.5 A historical review of bed widths.

Ware (*c.* 1580) was a little smaller than this, measuring 10 ft $8\frac{1}{2}$ in. wide by 11 ft 1 in. long (3265 × 3380 mm). It came from the Crown Inn, in the village of Ware in Hertfordshire (on the old road from London to Cambridge) where it was something of a tourist attraction. It is now in the Victoria & Albert Museum. Parsons (1972) recounts the story of a party of six couples who travelled up from London to use it for 'a frolick'.

People who suffer from back trouble are often advised to sleep in 'a hard bed'. Experience indicates that, not uncommonly, this advice turns out to be incorrect – and in some cases an excessively hard bed can make things worse rather than better. Norfolk (1993) reports a questionnaire survey of the advice that osteopaths give their patients concerning these matters. Of the osteopaths in his sample, 93% said they offered their patients advice about choosing a bed; although, interestingly enough, 83% also said that they would welcome more technical information on the ergonomics of bed design. Many were very critical of what Norfolk refers to as the 'current vogue' for excessively hard mattresses, with 98% of the sample saying that (presumably in their experiences and those of their patients) beds can be too firm for comfort. This has been confirmed in a user trial reported by Nicholson *et al.* (1985).

Part of the problem seems to be based upon a confounding of two different physical properties of the bed which we could call 'conformability' and 'sag'. Conformability is the ability of the bed to adapt to the contours of the body and to support it in a diversity of positions with a minimal build-up of pressure hot-spots. Conformability is principally a property of the mattress itself. There are a variety of ways of achieving this technically, in terms of the design of the bed springs, etc.

Unless the mattress is very soft indeed, however, the tendency of a bed to sag into a hammock shape will be more a property of the construction (or state of wear) of the supporting surface upon which the mattress is placed. For both sleeping comfort and postural support, it would seem desirable that the combination of mattress and bedstead (or supporting surface) should provide conformability without sag.

CHAPTER EIGHT

Health and safety at work

About 400 people are killed each year in the UK, in accidents that take place at work. (The exact figure fluctuates a little from year to year.) A further 16 000 are seriously injured; and at least ten times this number sustain injuries that although they are of a less severe nature, are none the less serious enough to keep them off work for three days or more (and thus find their way into the official statistics). Added to this we have an unknown (but doubtless very great) number who sustain minor injuries requiring first aid treatment only, and an unknown (but again large) number who develop diseases or ill health, of one sort or another, as a result of their work. (Figures from HSE Annual Reports.)

If we take fatalities as an index, however – and there seems to be good reason to do so, since they are likely to have been recorded more carefully than mishaps with less severe consequences – then the available evidence seems to indicate that work is getting steadily safer. The UK annual fatality rate currently stands at around 1.3–1.7 deaths per 100 000 employees. In 1981 it was 2.1 per 100 000; in 1971 it was 3.6 per 100 000; in 1961 it was 5.6 per 100 000; and in the first decade of this century it was 17.5 per 100 000 (Figure 8.1). The downward trend is thought to be due in part to better regulation of working practices, and in part to changes in the nature of work such that fewer people are engaged in its more hazardous varieties.

Table 8.1 shows fatal and non-fatal accidents broken down by their principal direct cause, as recorded in the official statistics. The figures for non-fatal accidents are for 1992 (the most recent statistical year available at the time of writing.) The figures for fatalities are based on the previous seven years, which have been lumped together, in order to average out the annual fluctuations which arise in the data because of the relatively small numbers involved. The fatality figures do not include the 167 lives that were lost on 6 July 1988 in the Piper Alpha oil rig disaster.

The table is set out by rank order for the non-fatal accidents. The data have a number of striking features. In the case of the non-fatal accidents, the top seven statistical categories account for about 90% of all the accidents. For the fatalities, the grouping is slightly less marked, with the top seven categories accounting for just 80% of the total. More importantly, the rank orderings for the fatal and non-

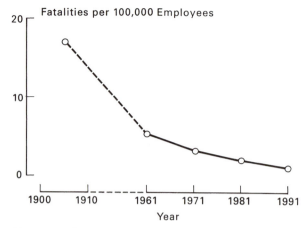

Figure 8.1   Fatal accidents at work, 1900–1991.

fatal accidents are quite different. The largest difference is for lifting and handling accidents, which take first place in the case of the non-fatal accidents, but are in last place in the case of the fatalities. You would find similar differences if you were to compare non-fatal accidents having different degrees of severity. The differences can in many cases be predicted on a common sense basis, in that they reflect the relative propensity for causing serious injury of different types of mishap. Thus the relative positions of 'fall on the level' and 'fall from a height' are different for the fatal and non-fatal accident in Table 8.1. Overall you are much more likely to fall on the level

Table 8.1   Accidents at work, classified by principal cause.

Principal cause	Non-fatal % Total	Rank	Fatal % Total	Rank
Lifting, handling, carrying	32.0	1	0.3	14
Slip, trip, fall on same level	21.4	2	1.2	11
Struck by moving object (incl. falling object)	13.8	3	13.2	3
Fall from a height	9.5	4	25.0	1
Striking against stationary object	6.0	5	0.6	12
Contact with moving machinery	4.9	6	7.5	5
Contact with harmful substance	2.9	7	2.4	8
Struck by moving vehicle	2.5	8	15.8	2
Electricity	0.5	9	6.2	6
Animal	0.5	10	0.3	13
Trapped by collapsing or overturning object	0.4	11	8.9	4
Fire	0.4	12	1.8	10
Explosion	<0.1	13	2.3	9
Drowning and asphyxiation	<0.1	14	4.0	7

Source: All figures based on HSE Annual Reports. Non-fatal accidents are for the year 1992. Fatal accidents are for the period 1986–1992 (excluding those resulting from the Piper Alpha disaster).

than to fall from a height; but if you fall from a height the injuries are more likely to be severe ones.

The relative frequencies with which accidents having consequences of varying degrees of severity occur, are often summarized in the form of an *accident triangle*. Figure 8.2 is an example, based on the UK figures summarized at the beginning of this chapter together with additional data from various sources. We note, however, that the shape of the triangle will be very different for different types of accident. Thus for 'fall from a height', where we have about 130 lost-time injuries per fatality, the triangle is sharply peaked; whereas for 'lifting and handling' accidents (about 50 000 injuries per fatality) the 'triangle' is almost flat.

The accident triangle is a reflection of the two factors (or sets of factors) that determine the risks inherent in an activity or working practice: the probability of a particular event (i.e. accident) occurring, and the probability of particular consequences (i.e. injury) resulting from such an event (should it occur).

The distinction can be an important one, insomuch as the steps required to control these two components of risk may in some cases be different. Thus the distinction is sometimes drawn between *primary safety*, the prevention of accidents *per se* and *secondary safety*, the protection of the person in the accident situation. (For example designing safer roads and more 'crashworthy' vehicles respectively; or making loads easier to handle, as against providing safety boots in case you drop them on your feet.) An equivalent distinction may be drawn for preventive medicine in general, where it is customary to speak of *primary prevention* (of the precursors of disease), *secondary prevention* (of the disease itself) and *tertiary prevention* (of its long-term consequences).

A closely related distinction is the one that is nowadays drawn between *risk* and *hazard*, a hazard being the potential to cause harm and a risk being the likelihood that this harm will be realized. (The UK Health and Safety Executive now uses these terms in this way.) The distinction is clearly an important one. The terminology is

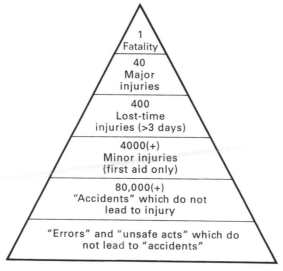

Figure 8.2 The accident triangle.

confusing, however, since in everyday language the words 'risk' and 'hazard' are used interchangeably (see, e.g., the definitions of these words given in the Oxford English Dictionary).

8.1   Human factors in industrial safety: an overview

An *accident* is an unplanned, unanticipated or uncontrolled event – generally one that has unhappy consequences.

There are two main theories as to the prevention of accidents:

- *Theory A*: Accidents are caused by unsafe behaviour; they may therefore be prevented by modifying the ways in which people behave.
- *Theory B*: Accidents are caused by unsafe systems of work; they may therefore be prevented by redesigning the working system.

The former approach could be characterized as 'fitting the person to the job (FPJ)'; the latter as 'fitting the job to the person (FJP)'. The two theories may be regarded as complementary. Neither provides a complete explanation of the ways in which accidents occur, but both tell us an important part of the story.

Unsafe behaviour may stem from:

- a lack of awareness of the risks of work;
- a foolhardy attitude towards the risks of work;
- lack of adequate instruction and training in safe working methods;
- lack of supervision in the working situation.

Not all unsafe acts lead to accidents; likewise, not all accidents result in injuries (see Figure 8.2). In general, unsafe behaviour is common and accidents are rare (and injuries more so). Unsafe behaviour is thus reinforced.

The basic elements of a *safe working system* are:

- a safe working environment;
- safe plant and equipment;
- safe procedures and working practices;
- competent personnel.

The employer has a responsibility at law (in the UK, both under the Health and Safety at Work Act and in terms of his common-law duty of care): *not only* to warn his employees about the risks of work, to provide adequate and sufficient instruction training in safe working methods, and to put the basic elements of a safe working system in place, *but also* to take such steps as are necessary to ensure that the system of work continues to operate safely on an ongoing basis. The latter (which is in many ways the more difficult of the two) is contingent on what is often called the *safety culture* of the organization: a set of factors, deeply embedded in its social ethos, that colour the attitudes of its members and influence their actions and risk-taking behaviour (at all levels in the organizational hierarchy).

Accidents commonly have multiple causes, in that they stem from the conjunc-tion (i.e. coming together) of a number of adverse circumstances. It is widely accepted that *human error* makes a significant contribution in the causation of very many accidents at work – probably the overwhelming majority. The contribution may very well be a decisive one, in that *but for* the error in question the accident would not have occurred. Errors do not arise in isolation, however. Their occurrence is very often contingent upon other adverse circumstances or features of the working system that lie outside the jurisdiction and control of the person concerned.

Psychologists of the cognitive persuasion have made a number of attempts at classifying human error, doubtless to their own satisfaction. For practical purposes it is important to recognize two particular categories:

- errors of judgement in the appraisal of risk;
- errors of execution in the performance of the working task.

True errors of judgement in the appraisal of risk stand at one end of an unbroken continuum which stretches through violations of safe working practice (having greater or lesser degrees of conscious intent) to the deliberate and premeditated criminal act (of vandalism, sabotage, assault, etc.). Errors arising in task performance are commonly *system-induced*, in that there may be deficiencies in the design of the working system (most typically at the operator/machine interface) which make the person's working task more difficult and thus render him more error-prone. Notwithstanding that this may be so, however, such errors may also have an attitudinal component, in that by the investment of additional care and effort (conscious or otherwise) it may in some cases be possible for the person to adapt to (and cope with) the deficiencies of the system. So looked at from another standpoint, the system-induced performance error may be construed as stemming from a want of attention, etc.

In English law, a *crime* has two components: the criminal act (*actus rea*) and the criminal intent (*mens rea*). The law recognizes the existence of an important grey area, between the true error (for which no blame accrues) and the premeditated criminal act. This recognition forms the basis for the concept of *negligence*: the failure to act with a reasonable degree of caution or prudence in the face of a risk that is foreseeable 'in reasonable contemplation'.

At law, an injury may stem from a 'true accident' (for which no one is to blame); it may stem from the negligence of a single party (e.g. either the employer or the employee); or it may be attributable, in various measures, to the negligence of two or more parties. Thus the outcome of a personal injury claim might, for example, be that the losses arising from that injury were attributable principally to the negligence of the employer (in failing to institute a safe system of work); but that there was also a significant element of 'contributory negligence' on the part of the employee (in failing to take reasonable care for his own safety). The damages awarded would be adjusted accordingly – on the basis of the court's estimate of the relative magnitude of the two causative contributions. In other words, theories A and B of accident causation would both apply in part.

8.1.1 The catastrophic failure of complex systems

When an accident has particularly serious consequences (e.g., multiple loss of life or large-scale environmental contamination), we are likely to refer to it as a 'disaster' or a 'catastrophe'. I have dealt at length elsewhere with the role of human error in the catastrophic failure of large-scale human-made systems – nuclear incidents, plane crashes, etc. (Pheasant 1988a, 1988b, 1991a). For present purposes we shall limit ourselves to just one example, the loss of the *Herald of Free Enterprise*, which illustrates the main points at issue particularly well.

The reader will recall that on 6 March 1989, the cross-Channel car ferry *Herald of Free Enterprise* put to sea from Zeebrugge with her bow doors open. An inrush of water flooded the large unobstructed spaces of her lower car deck, causing her to capsize, and 188 lives were lost. The direct responsibility for ensuring that the bow doors were closed lay with the Second Mate who, as it transpired, was asleep in his cabin (where he remained until he was awakened by the ship rolling over). The overall responsibility for the safety of the ship lay with the Captain, who was in his normal place on the bridge. From a human factors standpoint the most striking feature of the accident was that from his customary position on the bridge, the Captain had no direct means of knowing whether the bow doors were open or not. There was no visual display – such as a simple warning light, for example – to provide him with this critical information. Neither was it anyone's particular duty to tell him (although this latter point remains surrounded by a certain air of vagueness). The Captain's fatal decision to put to sea may thus be construed as a classic system-induced error.

A number of other adverse circumstances were contributory factors. The tide that day was particularly high, making the loading of the ship difficult (particularly since

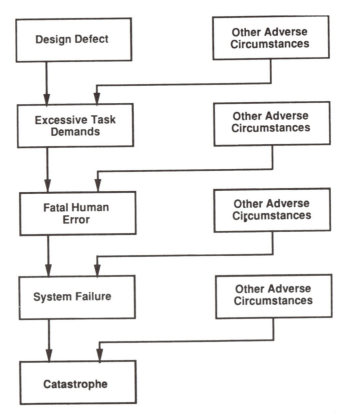

Figure 8.3 The Zeebrugge–Harrisburg syndrome (*New Scientist*, 21 January, 55–58 from S. Pheasant, 1988)

the ramp at Zeebrugge had been designed for a different type of vessel); because of the fierce economic competition of the cross-Channel ferry routes, the crew were under considerable pressure to achieve the fastest possible turnaround time in port; because of vandalism, life jackets were stored in inaccessible lockers; and so on. The relationship between these various contributory factors, in the chain of causation which led to the catastrophe, are summarized in Figure 8.3. I have referred to this overall process as the *Zeebrugge–Harrisburg Syndrome*. (The reader will recall that the ill-fated Three Mile Island nuclear reactor was in Harrisburg, Pennsylvania.)

It became apparent at the subsequent Court of Inquiry that it was by no means unknown for ferries of this type to go to sea with their bow doors open. Other captains had commented on the problem and the suggestion that warning lights should be installed had been passed up to the board level of the ferry company. The suggestion met with derision. The honourable Mr Justice Sheen concluded that 'from top to bottom the body corporate was infected with the disease of sloppiness'. The Captain, the First Officer and the Second Mate were found to be negligent. The Captain lost his operating certificate for one year and the First Officer for two years.

The episode has two interesting legal postscripts. The Captain appealed against the loss of his certificate on the grounds that going to sea with the bow doors open was a common practice. The appeal failed on the grounds that the fact that a particular form of negligence was rife in the world of car ferries did not condone it in any individual case. This finding is in some measure unusual in that the defence of 'normal custom and practice' is often successful in personal injury claims. An attempt to bring a criminal prosecution against the ferry operators for 'corporate manslaughter' also failed, it being ruled that the risks attendant on going to sea with the bow doors open were not sufficiently obvious to warrant such a charge. The common-law test of 'reasonable foreseeability' does not apply here, the criterion applied in the criminal charge of manslaughter being a more demanding one. The risk would have to 'stare you in the face'.

8.1.2 Everyday accidents

We now return to the everyday accidents of the shop floor as categorized in Table 8.1. Each category of these presents its own set of ergonomics or 'human factors' issues.

Slipping, tripping, and falling accidents (for example) very often result from a lack of 'good housekeeping' – the failure to mop up spillages, keep walkways free from obstructions, trailing cables, and so on. This stems in turn from a defective safety culture and 'the disease of sloppiness'. Issues of environmental design may also be involved, however, for example the layout and lighting of the working area, the slip resistance of flooring materials, etc. This author continues to be amazed by the unsuitability of the flooring materials used in public buildings – especially around entrances, etc., where in bad weather the floor can get muddy and wet faster than staff can be reasonably expected to keep it clean and dry.

Contact with machinery accidents remain a common cause of serious injury at work. The safeguarding of machinery raises some interesting points in the theory of anthropometrics. The concept of a *safety distance* is based upon a reversal of the normal criteria of reach and clearance. A safety guard or barrier will fulfil its function of separating people from the hazardous parts of machines, *either* if apertures in the guard are sufficiently small to prevent access by a particular body part (finger, hand, arm, etc.), *or* if the distance between the aperture and the hazard is sufficiently great for the latter to be out of reach (by the body part in question). The limiting user is thus one with a small finger (hand, arm, etc.) in the case of aperture size and a long finger (hand, arm, etc.) in the case of the distance. (In theory we also need to allow for the correlation between the length and girth of the body parts in question; but in practice this is likely to be small and if we assume it to be zero we will err on the side of caution.) Safety distances are the subject of a series of British and European Standards to which the reader is referred for further information.

Regrettably it is all too common for people to seek ways of 'defeating' the safety mechanisms of machinery in the interests of increased output – and for serious or fatal injury to result. There have been prosecutions under the UK Health and Safety at Work Act for fatalities caused in this way.

The reader will thus note that for both the classes of accident discussed above, theories A and B are both applicable to some extent as to accident causation. Overall, this is true for most other classes of accident too – up to and including the catastrophic failure of complex human-made systems.

We note also the very general applicability of 'the ergonomic approach' to accident prevention. We turn now to a large and important class of work injuries in which ergonomic issues are of decisive causative significance and in which theory B will in general be very much more applicable than theory A.

8.2   Ergonomic injuries

An *ergonomic injury* is one that occurs as a direct or indirect consequence of the nature and demands of the person's working task, rather than as a result of some hazard to which the person is exposed, during the course of his or her work, but which is not intrinsically part of the working task itself.

Ergonomic injuries include (for example):

- lifting and handling injuries;
- work-related upper limb disorders;
- musculoskeletal pain and dysfunction resulting from unsatisfactory working posture, etc.

Ergonomic injuries do not include those resulting from toxic or environmental hazards to which the person is exposed at work, although environmental factors (e.g. heat, cold, etc.) may play a contributory role in the causation of ergonomic injuries.

In other words, ergonomic injuries result from a *mismatch* between the *demands* of the working task and the *capacity* of the working person to meet those demands; generally when the former exceeds the latter and the person is placed in a situation of *overload*.

Ergonomic injuries may occur as discrete events which take place at a particular point in time as a result of a single episode of *over-exertion*. They may occur insidiously over a period of time as a result of *cumulative over-use*. Or they may result from a combination of both, in that the effects of cumulative over-use may render the person susceptible to subsequent injury by over-exertion.

Over-exertion injury occurs when an anatomical structure fails under peak loading, because its mechanical strength (usually its tensile strength) is exceeded. This most characteristically occurs in the execution of some voluntary action.

Over-use injury occurs when the rate of damage to an anatomical structure exceeds the rate of repair. The injury process usually involves repeated micro-trauma. The repair of damaged tissue is a natural ongoing biological process. The timescale may be one of hours, days, months, or years.

Cumulative trauma to soft tissues (and other anatomical structures) resulting from prolonged over-use may lead to a progressive diminution in their mechanical strength, thus rendering the structure in question more susceptible to injury by over-exertion at some subsequent point of peak loading (perhaps at a level of loading that under other circumstances, could be tolerated with impunity). This 'history dependence' of tissue strength when exposed to repeated cycles of loading is sometimes referred to as a 'creep effect'; it may be likened to the phenomenon of metal fatigue. The physiological fatigue of muscles may also be a factor in that it may lead to a breakdown in the normal co-ordination and control of voluntary movement, with the attendant risk that an articulation will be driven beyond the limits of its normal range of motion.

Some ergonomic injuries are 'accidents' in the technical sense of the word as defined above. For example, a person may lose his balance when handling a load that is beyond his safe capacity – and overstrain himself in trying to regain control of the situation. But the great majority of over-exertion injuries are not accidents in this sense of the word – in that they do not entail any identifiable intervening event that interrupts the normal execution of the action in question – other than the direct manifestation of the injury itself. Thus the person may feel a 'sudden sharp pain' (in his back, shoulder, wrist, etc.) whilst executing a familiar action or procedure in what appears to be the normal way. The only thing that is unexpected is the pain itself.

There is also another and more radical sense in which these injuries are not 'accidents' – and that is that they are very often entirely foreseeable.

Consider, for example, a working population such as nurses, who are called upon to handle the awkward, unstable and excessively heavy load of the human body, on a frequent and repeated basis, and very often under circumstances that are adverse in other respects. It is wholly predictable that, as a population, nurses will suffer a high incidence of back injuries. We can also predict (both on a theoretical bio-mechanical basis and on the basis of experience) the kinds of patient-handling manoeuvres in which such injuries are most likely to occur, although we may not be able to predict (with any degree of accuracy) whether a particular nurse will injure herself on a particular occasion (see Pheasant and Stubbs 1992a).

We could say much the same thing for the over-use injuries to the hand, wrist, and arm, which are endemic in people who do repetitive hand-intensive work on industrial assembly lines. It has been known for over forty years that people who work in motor-car assembly plants are liable to suffer from these sorts of conditions (Thompson *et al.* 1951). The risk is equally well recognized in the poultry processing industry, to the extent that, in these industries at least, such injuries must be regard-ed as 'foreseeable' in any conceivable meaning of the word.

The back injuries of the nurse or the upper limb disorders of the assembly-line worker are inherent in the very nature of the systems of work in question. The employer is under a legal obligation to take such steps as are 'reasonably practi-cable' to set up a safe system of work. Very often he does nothing of the kind.

8.3   Back injury at work

At the present time something in the order of 70 million working days are lost each
year in the UK due to back pain. The figure is growing rapidly; it is growing on an
upwardly accelerating curve; and it is growing more rapidly in women than in men.
We do not know why.

Back pain is a condition of complex multifactorial aetiology in which many risk
factors may play a part. These may be grouped under two main headings:

- *occupational (or ergonomic) risk factors*, associated with the mechanical stresses
  to which the person is exposed in the course of his or her working life;
- *personal risk factors*, particular to the individual concerned, which may be further
  subdivided into those factors associated with the person's lifestyle and those that
  stem from his or her constitutional make-up.

The epidemiological literature dealing with these matters is vast. Taken as a whole it
indicates that (in a statistical sense at least) occupational risk factors have a greater
causative significance than personal risk factors; and that of the personal risk
factors, the lifestyle factors are of greater causative significance than the constitutional factors (see Pheasant 1991a). One would note, however, that since these
factors must almost certainly act interactively and in complex combinations, the
picture in any individual case may be very much less clear cut.

Although now more than twenty years old, the classic epidemiological studies of
the Israeli researcher Magora (1972, 1973a, 1973b) remain in many ways definitive.
These were based on a large sample of men and women drawn from many different
walks of life. To summarize the results of what was a very extensive investigation
indeed, Magora's findings indicate that two quite distinct groups of people are particularly at risk: those whose jobs were physically very demanding (entailing frequent heavy lifting, forceful exertion, etc.); and those whose jobs were fully
sedentary. But those falling into a middle category, whose jobs were moderately
physically demanding and who spent some of their time sitting and some of their
time standing, fell into a particularly low-risk category. Psychological factors were
also important, in that people who reported low levels of job satisfaction, or found

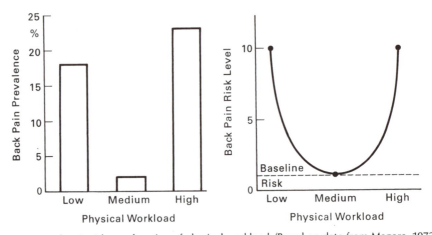

Figure 8.4 Back pain risk as a function of physical workload. (Based on data from Magora, 1972.)

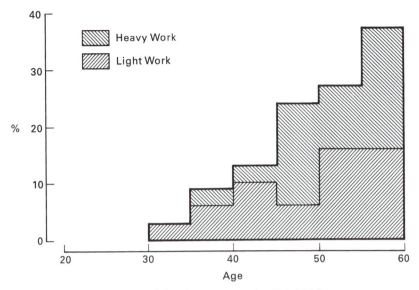

Figure 8.5 Prevalence of pronounced disc degeneration, after Hult (1954).

their work mentally demanding (in the sense of the demands it placed on their concentration), were more likely to suffer with their backs.

The most striking feature of Magora's findings was the magnitude of the difference between the categories. About 20% of people in the physically demanding and fully sedentary categories suffered with their backs, as against around 2% in the middle low-risk category. In other words there was a ten-to-one difference in risk level. The inference to be drawn is that the relationship between back pain risk and physical workload is U-shaped (or possibly J-shaped) – see Figure 8.4. Let us suppose that the relatively low prevalence in the middle category represents a baseline level of risk attributable in a general sense to 'the human condition' (or, to the summative effects of all the various personal risk factors). It would follow that any excess prevalence, over and above this level, found in the other categories, represents back trouble in which the person's work was a significant causative factor. In the great majority of cases therefore, back trouble must be regarded as a work-related condition. In other words, back trouble is an ergonomic injury.

Ergonomic risk factors for low back trouble are summarized in Table 8.2 (see Pheasant 1991a for a summary of the literature upon which this table is based).

The probability is that these various factors have an additive effect. There is actually surprisingly little direct evidence for this proposition – although it certainly

Table 8.2 Back pain: ergonomic risk factors.

- Heavy work – lifting, pushing, pulling, sudden maximal force exertion, bending, twisting, etc.
- Stooped working posture
- Prolonged sedentary work
- Lack of task diversity
- Unaccustomed physical activity
- Vibration and shock
- Psychosocial factors

makes physiological sense – and given that the different risk factors are not associated with different sorts of clinical problems (and we have no particular evidence
that this might be the case) then it is difficult to see how things could be otherwise.

The probability also is that these factors act cumulatively over a period of time
(although the process of cumulative injury will doubtless be offset to some extent by
the body's natural mechanisms of repair). There is some very striking epidemiological evidence for this – at least over the very long time scale. Degenerative disc
disease is generally thought of as being part of the natural ageing process. The
physiological changes in the properties of the disc which underly the process of
degeneration commence at around age 25 or so; and past middle age we are all
affected to a greater or lesser extent. Superimposed over these physiological changes
are the effects on the discs of 'normal wear and tear'. A constitutional predisposition
is also thought to be involved, although the principal evidence cited for this proposition is that people who show severe signs of degenerative changes in one part of the
spine are likely to show them in other parts as well, and this evidence could be
interpreted in other ways. As part of a much larger epidemiological study of back
and neck trouble, Hult 1954 compared the prevalence rates of radiological signs
indicative of advanced disc degeneration, in men who were in light and heavy
occupations respectively. The results are shown in Figure 8.5. In the youngest age
group there is no detectable difference in prevalence. With the passing of the years
the prevalence for the two occupational categories diverge, until in the oldest (>50)
age group there is a difference in prevalence of around two to one. In other words,
in addition to the 'normal wear and tear' of everyday life, the backs of men in heavy
jobs showed clear and objective signs of a further degree of 'abnormal wear and
tear', resulting from the nature of their work.

We shall turn now to the particular ergonomic problems associated with the
lifting and handling of heavy loads at work.

8.4 Lifting and Handling

Recent decades have seen major changes in the nature of industrial work as human
muscle power has been increasingly replaced by machines. Overall, work is not as
heavy as it was forty years ago. Lifting and handling injuries continue to be a major
problem, however. The percentage of all reported work injuries attributed to lifting
and handling has not shown much change since the early 1950s. This is something
of a paradox.

The percentage of injuries attributable to lifting and handling in different sectors
of the economy are likewise surprisingly similar. The figure is a little lower than
average in banking and finance and a little higher in the construction industry, but
the differences only amount to a few per cent either way (HSE 1992b).

The one area of working life that stands out is the healthcare services where
lifting and handling accounts for over half (55%) of all reported injuries leading to
three days' absence from work, as compared with around one-third (32–34%) for the
working population as a whole. The difference is attributable to the particular difficulties attendant on lifting and handling the human load. Even within this high-risk
sector, however, there are also striking differences between occupational groups.
Ambulance personnel have a very much higher incidence of patient-handling
injuries than nursing staff; and nursing auxiliaries, student nurses and community

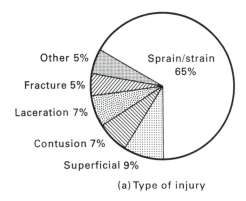

Other 5%

Sprain/strain
65%

Fracture 5%

Laceration 7%

Contusion 7%

Superficial 9%

(a) Type of injury

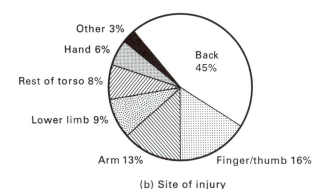

Other 3%

Hand 6%

Back
45%

Rest of torso 8%

Lower limb 9%

Arm 13%

Finger/thumb 16%

(b) Site of injury

Figure 8.6 Lifting and handling injuries classified (a) by type and (b) by site.

nurses all have a higher injury rate than qualified nurses working on the wards (HSE 1982, Pheasant and Stubbs 1992a). In general these differences reflect differences in the overall amount of lifting which the occupational groups in question are called upon to do and the difficulty of the circumstances under which they are called upon to do it.

Discussions of the prevention of lifting injuries have tended to revolve around two seemingly simple questions:

- What is the safest way of lifting heavy weights?
- What is the maximum safe weight a person can lift?

The first question stems from theory A as to accident causation, the second from theory B. Neither has a simple answer.

The available evidence points to the conclusion that training people in 'safe' lifting techniques alone is unlikely to have a sustained impact on injury rates. This is equally so both for lifting training in general and for the special case of patient handling. Training is necessary but not sufficient. I have discussed the matter of lifting training at length elsewhere (Pheasant 1991a, Pheasant and Stubbs 1992b). The following discussion will concentrate on the design of safe systems of work.

Not all lifting injuries are over-exertion injuries: a significant minority are accidents in the narrow sense of the word (as defined above) and result in lacerations, contusions, fractures, and so on. Neither do all lifting injuries affect the back,

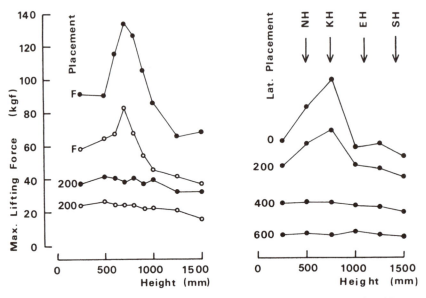

Figure 8.7   Strength of a static lifting action as a function of height above ground and foot placement. Left: freestyle placement (F) and feet placed behind the axis of lift. Data kindly supplied by Anne-Marie Potts. ● = 16 men; ○ = 14 women. Right: feet placed 20 mm behind the axis lift 400 mm apart, various distances to the left. The placement figures are for the mid-line of the body. NH = knee height, KH = knuckle height, EH = elbow height, SH = shoulder height of the 21 male subjects. Data kindly supplied by Jane Dillon of the Furniture Industry Research Association.

although the low back is the part of the body most commonly affected (see Figure 8.6). It is worth noting in particular that lifting tasks not infrequently result in over-exertion injuries to the neck, shoulder and wrist.

We have noted already that heavy manual work may accelerate the degenerative processes that occur in the spine with age. There is also good epidemiological evidence for an association between work that entails heavy lifting in a squatting or kneeling position and osteoarthritis of the knees (Cooper *et al.* 1994).

In summary, lifting and handling tasks entail three distinct classes of risk:

- the risk of accidental injury;
- the risk of injury due to over-exertion;
- the risk of injury due to cumulative over-use.

In practice, however, the measures that are required to control these three classes of risk will tend to be similar ones.

8.4.1   Workspace layout

The first and most fundamental principle of safe lifting is that the load should at all times be as close as possible to the body. There are two reasons for this. First, the closer the load, the less is its leverage about the various articulations of the body; hence less muscular effort is required and there is less mechanical stress on potentially vulnerable structures (e.g. those of the back). Second, the closer the load, the more easily it is counterbalanced by the weight of the body so it is less likely to get out of control. Thus the strength of the lifting action falls off rapidly as a function of the distance of the load from the body – and the weight that can be handled safely becomes correspondingly less (see Figure 8.7).

A second important principle is that symmetrical lifting actions are in general safer than asymmetrical lifting actions, particularly if the latter involve turning actions which impose a rotational twist on the spine. This is partly because the lumbar spine is anatomically vulnerable to injury under torsional loading; and partly because in turning actions we naturally tend to 'lead with the hips', thus exposing the lumbar spine and its musculature to particularly high levels of loading.

In practice the distance of the load from the body and the symmetry of the lifting action will be largely determined by foot placement – and this in turn is determined by the presence or absence of obstacles that prevent the person from getting his feet beneath or around the load (Figure 8.8). Lifting and turning actions likewise very often stem from deficiencies in workstation layout.

Given good foot placement, the strength of the lifting action is greatest at around knuckle height (c. 700–800 mm) and falls off rapidly above and below this level. When exerting a vertical lifting force at knuckle height or thereabouts, the upper limbs are vertical and almost straight and the hips and knees are slightly flexed. The muscles of the lower limb thus exert a powerful extensor thrust along the line of the almost straight limb at their best possible mechanical advantage. When the force is exerted at a distance from the body, however, this peak in lifting strength disappears (see Figure 8.7).

If the lift commences at much below knuckle height, the person will either have to incline his trunk and therefore increase the loading on his spine (which will tend also to be flexed upon itself and thus anatomically vulnerable to injury), or else

Figure 8.8 Lifting at a distance: palletization task. From an original in the author's collection. (From S. Pheasant, *Ergonomics, Work and Health*, Macmillan, 1991, fig. 15.17, p. 302, reproduced with kind permission.)

strongly flex the knees, thus reducing the mechanical advantage and also rendering the knees anatomically vulnerable to injury. In either case the power of the lifting action is diminished and the weight that can be handled safely is likewise reduced.

A lift that is commenced at knuckle height or lower can (provided the load is not excessive) be continued comfortably to elbow height or a little more. If the load is a box or carton which is held by its lower edges, or a crate with handholds on the sides, the lifter will then begin to encounter difficulties as his wrist reaches the limit if its range of abduction. He will thus either have to change his grip or else make awkward compensatory movements of his upper limbs and trunk, neither of which is at all desirable (see Figure 8.9). The wrist is also anatomically vulnerable in this position.

Lifts that commence at elbow height may be continued to shoulder height or thereabouts without too much difficulty, but beyond that point the reduction in strength really begins to tell. There is a particular danger that loads that must be handled at shoulder height and above will get out of control. In this author's experi-

Figure 8.9   Lifting outside the normal height range. Note the hyperextension of the lumbar spine.

ence, lifting tasks that entail the handling of loads outside the comfortable height range of the person in question are a very common cause of injury.

On the basis of these considerations we may divide the reach envelope of the standing person into *lifting zones*, as shown in Figure 8.10 (after Pheasant 1991a; Pheasant and Stubbs 1992b). The heights given for the various landmarks are based upon anthropometric data for the 'standard reference population' (see section 2.4) but rounded up to convenient whole numbers. The verbal categories describing each zone may be regarded as giving a general indication of the weight of load that might be considered acceptable in each zone (see also below).

When carrying a load such as a box or carton, the person will generally hold it by the lower edges at hip height or above (800–1100 mm) so as not to impede walking. The effort required to lift loads from conveyor belts, etc., may often be reduced therefore by setting the belt at a level that allows the person to pull the load towards him and take its weight at a suitable height for carrying. (This will depend in some measure on the nature of the load.)

Pushing and pulling actions are generally performed most easily at between shoulder height and elbow height or a little below, depending on the circumstances. According to biomechanical studies by Ayoub and McDaniel (1973), the optimum

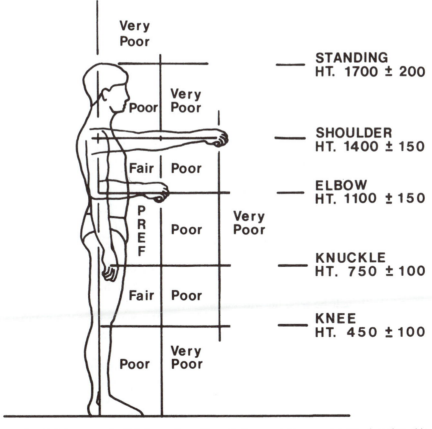

Figure 8.10 Height ranges for lighting actions. (From S. Pheasant, *Ergonomics, Work and Health*, Macmillan, 1991, fig. 15.20, p. 305, reproduced with kind permission.)

Table 8.3   Recommendations for the design of storage, shelving, and racks[a]

Height (mm)	Application and comments
<600	Reserve storage for rarely required items. Fair accessibility for light objects; poor for heavy
600–800	Fair for heavy items; good for light items
800–1100	OPTIMAL ZONE FOR STORAGE
1100–1400	Fair–good for light items–visibility unimpeded, access fair. Poor for heavy items
1400–1700	Limited visibility and accessibility. Most men and women will be able to stow and retrieve light items (at least on the edge of the shelves)
1700–2200	Very limited access; beyond useful reach for some people
>2200	Out of reach for everybody

Notes: [a] For present purposes, loads greater than 10 kg are deemed 'heavy'. Shelf depth should not exceed: 600 mm at heights of 800–1400 mm; 450 mm at heights less than 800 mm; 300 mm at heights greater than 1400 mm. Minimum acceptable unobstructed space in front of the shelves: 680 mm for small items at heights greater than 600 mm; 1000 mm at heights less than 600 mm or for bulky items at any height.

level is 70–80% of shoulder height – which works out at a little below elbow height or about 1000 mm for men and 900 mm for women. Fixed horizontal handles on trolleys, carts, etc., should be at this level; but vertical handles will often be a better solution in that they allow the user to find his or her own level.

Pushing actions are strongest when the feet are placed as far back as possible; pulling actions when the feet are as far forward as possible. High-friction shoes and flooring materials are important. An unobstructed floor space of 1000 mm is required; 1800 mm is preferable for pulling actions.

Tasks involving the storage of items on shelving and racks constitute an important class of handling problems. In general the heaviest and/or most commonly used items should be stored in the most accessible positions (see Section 5.1). Table 8.3 provides some guidance in these matters (based on the above anthropometric considerations and also user trials reported by Thompson and Booth 1982).

8.4.2   The load

In general a compact load is safer to lift than a bulky load of the same weight because its centre of gravity will be closer to the person's body; so the movement about his back will be less, he will be more stable on his feet, and so on. This is especially important if the load is to be lifted from the ground, since a compact load (<300 mm wide or from back to front) can be lifted between the knees rather than in front or to the side of the knees.

Unstable loads and loads with unexpected inertial characteristics are a particular hazard. The centre of gravity should be as close as possible to the geometrical centre; and if offset, its position should be marked. Contents should be securely packed to prevent unexpected shifts in the centre of gravity – these are a common cause of lifting injuries.

Secure handholds are an advantage (see Chapter 5).

8.4.3 Weight limits

What is the maximum weight that a person of a particular age and sex, and of 'normal fitness', may be expected to lift under a given set of circumstances, without undue risk of injury?

This is a very difficult question indeed. First of all, there is no weight of load, however small, that guarantees safety.

You can injure your back by stooping down to pick up a pencil. Biomechanical calculations show that when the trunk is inclined forward to a horizontal position, the loading on the base of the spine is the same as the loading that results from lifting a compact 30 kg weight close to the body (Pheasant 1991a). So if for one reason or another the back is at all vulnerable to injury there may be no spare capacity for actually lifting an external load, over and above the weight of the body itself. Unfortunately, the *precursors of lifting injury*, which determine a person's level of vulnerability, are not always easy to recognize. A person may be vulnerable to injury without knowing it. Overall we are not particularly good judges of what we can handle safely – and we commonly injure ourselves lifting loads that we believe to be within our capacity. Conversely, experience shows that if a load *feels* as if it is too heavy to handle safely then it very probably is too heavy. In other words, our subjective appraisals of our safe limits are systematically biased in the direction of risk (although there are doubtless important individual differences in this respect).

In any given set of circumstances (as defined by load characteristics, lifting position, and so on), we should in general expect the risk of injury due to over-exertion, *for a particular individual*, to increase steadily with the weight of the load. The relationship may or may not be linear (but biological systems being what they are, a non-linear relationship seems the more probable). Given the nature of human variability, then the rate at which the risk of injury increases with the weight of the load will necessarily vary greatly between individuals. So if there are important *threshold effects* (where the level of risk takes a sharp upswing) then the location of these will likewise vary greatly. Taking one consideration with another, one would expect the risk of injury by over-exertion, *for a given working population*, to increase with load weight in a smooth upwardly accelerating curve, in which such threshold effects as might be present for given individuals are masked.

We also need to take into account *both* the immediate risks of over-exertion injury *and* the long-term risks of cumulative over-use – and probably also the interactions between the two. The risk of accidental injury presents an even more difficult set of problems. One would expect the probability of some unexpected mischance that leads to injury to increase with the weight of the load; but the connection is not as clear as it is for over-exertion injury or cumulative over-use, except insomuch as the heavier the load, the more likely you are to injure yourself seriously if it gets out of control.

Even if we fully understood these matters, to the extent of being able to plot out a curve relating overall population risk to load weight, we should still be faced with the question of where to set the limit. What is the cut-off point beyond which the risk of injury becomes unacceptable?

One possibility is to set the limit at a level of loading that would result in a *just noticeable risk of injury*: that is, at which the risk of injury due to work would be just measurably greater than the background level of risk associated with life as a whole. This is, broadly speaking, what we attempt to do for chemical hazards, radiation hazards, and so on, since in these cases the risk in question may be controlled by a more effective containment of the hazard so that the working person does not come into contact with it. In the case of lifting and handling, however, we cannot do this – and to set the limit at the level of just noticeable risk would to all intents and purposes be equivalent to calling for the abolition of all useful manual work. This would be pointless – not least because guidelines that cannot be met in practice are ignored and fall into disrepute.

Neither does it make much sense to set the limit at a level of loading at which injury becomes 'probable' to the extent of being 'more likely than not', because in practice this would mean that we were constantly having to replace our workforce. There are industries where this happens. They are recognizable by the age distributions of their workforces. The ambulance service is a notable case in point.

The problem thus becomes one of reaching a *reasonably practicable* compromise position which allows working life to continue without incurring excessive risk. This roughly approximates to the legal concept of a risk that is *reasonably foreseeable*; or, in this context, the level of risk which a 'reasonable person' would agree to accept in the course of his or her working life (were he or she fully appraised of the facts of the matter).

In reaching this point we are faced with two principal difficulties: the first is the inadequacy of our scientific knowledge; the second is that legal conceptions of probability are not altogether the same as scientific ones, in that 'moral certainty' does not equate easily with statistical certainty.

Some countries in the world have seen fit to impose limits on the weights that people may lift at work; others have not. The International Labour Office has published a compilation of such weight limits (ILO 1990). These are summarized in the form of a cumulative distribution in Figure 8.11. The graphs show the percentage of countries in which a load exceeding the weight in question would be considered unacceptable. For adult men these cluster around a median figure of 50 kg, with 50% of values falling within the 45–55 kg range; for women they cluster around a median value of 25 kg, with 50% falling within the 20–25 kg range.

The US National Institute of Occupational Safety and Health has published an influential set of guidelines which deal specifically with symmetrical two-handed lifting actions performed directly in front of the body (NIOSH 1981). For any such action it is possible to calculate an *action limit* (AL), beyond which there is deemed to be a moderate increase in risk; and a *maximum permissible limit* (MPL), beyond which the risk is considered unacceptable. These guidelines are based upon biomechanical, physiological, psychophysical and epidemiological criteria. The equation that defines the AL and MPL takes into account the horizontal and vertical positions of the load, the distance it is lifted, the frequency of lift, and the duration of the task. The original guidelines have subsequently been revised (Waters *et al.* 1993). The equation defining the limits has been modified, and two new elements, dealing with asymmetry and ease of grasp, have been added. The action limit and maximum permissible limit have been dropped in favour of a *recommended weight limit* (RWL) which is set at a level that is approximately equivalent to the old action limit; and the concept of a *lifting index* (LI) has been introduced. This is the ratio of the load on the job to the RWL. It thus represents a relative measure of the severity of risk. It is essentially seen as a tool in job redesign and no specific cut-off point is proposed. I have discussed the theoretical and practical strengths and weaknesses of the NIOSH guidelines at length elsewhere (Pheasant 1991a). I would further add

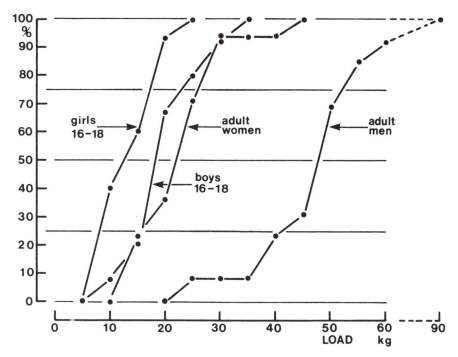

Figure 8.11 Statutory limits on the weight to be handled by one worker in different countries of the world. Cumulative distributions are based on data given in ILO (1990). (From S. Pheasant, *Ergonomics, Work and Health*, Macmillan, 1991, fig. 15.26, p. 314, reproduced with kind permission.)

that in my view both the old and new NIOSH equations greatly underestimate the importance of the vertical height range of the lift (see above).

The set of guidelines that was devised at the Robens Institute of Health and Safety at the University of Surrey was based upon an extensive series of laboratory experiments and field studies in which intra-abdominal pressure (IAP) was used as an indirect index of spinal loading. On the basis of the field studies an IAP of 90 mm Hg was adopted as the safe limit for men (Davis and Stubbs 1977, 1978); and an equivalent level of 45 mm Hg was subsequently adopted for women (David 1987). The Robens Institute figures have been incorporated into a UK Ministry of Defence Standard (Ministry of Defence 1984). The guidelines are presented in the form of contour maps, drawn in elevation and plane – examples of which are shown in Figure 8.12. The contour maps represent the level of loading at which the IAP criterion will be violated on only 5% of occasions. Similar maps are provided for pushing and pulling actions and correction factors are given for age and sex.

The 'guideline figures' published by the UK Health and Safety Executive (HSE 1992), and shown in Figure 8.13, are based in part upon the Robens Institute load limits and in part upon the concept of lifting zones as set out above. The document in which these are set out stresses that they are not 'limits' as such, but are for guidance purposes only – to be used in the context of a broader approach to the assessment of risk based upon ergonomic principles. The guideline figures are said to be such as to afford 'reasonable protection' to nearly all (95%) of men and between one-half and two-thirds of women. A correction factor of one-third is sug-

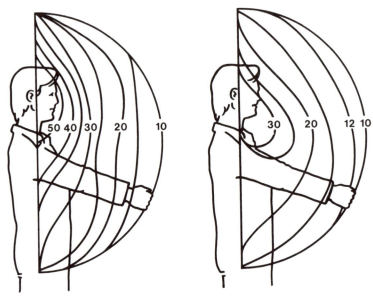

Figure 8.12 Suggested limits for lifting forces (kgf) two-handed (left) and one-handed (right) at various positions in the zone of convenient reach (ZCR) as given by Davis and Stubbs (1977, 1978) and Ministry of Defence (1984). Values given are for men under 50 lifting less than once per minute.

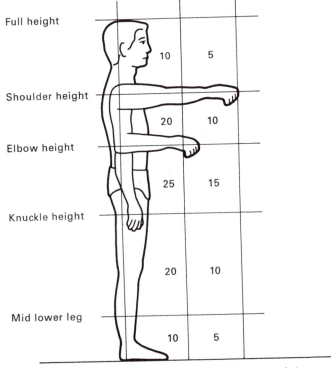

Figure 8.13 Lifting zones with the HSE guidelines data (measurements in kg).

gested as giving the same degree of protection to nearly all women. Correction factors are also given for twisting, repetitions rates and team lifting.

In essence, the guideline figures define a boundary level beyond which there is a potential risk of injury. The HSE document in question goes on to say that operations exceeding the guideline figure by a factor of more than about 2 should 'come under very close scrutiny' – presumably in terms of risk assessment. I take this to be a way of saying that beyond this point the risk of injury may well be unacceptably high.

8.5 Work-related upper limb disorders

The terms 'work-related upper limb disorder' (WRULD) and 'repetitive strain injury' (RSI) are, to all intents and purposes, synonymous. Both terms are used generically to refer to a diverse class of conditions affecting various anatomical sites in the hand, arm, shoulder, and neck, which occur in people doing a wide variety of types of work involving intensive use of the hands (not all of which are necessarily repetitive in the strict and narrow sense of the word).

Occupational groups most notably affected include:

- industrial assembly line workers, for example in the automotive, electronics, pottery and food processing (especially meat and poultry) industries;
- workers at supermarket checkouts;
- musicians (particularly those playing string instruments and the piano, but others as well);
- keyboard users (particularly data entry workers, copy typists, legal secretaries and journalists).

Table 8.4 summarizes the results of a number of epidemiological studies of disorders falling into this category, which were gathered together by Armstrong *et al.* (1993). Note that the 'relative risk' is the ratio of the frequency with which the

Table 8.4 Work-related upper limb disorders: relative risk in selected occupational groups.

Job	Risk
Industrial workers (high force/high rep.)	29.4
Sausage makers	24.0
Shipyard welders	13.0
Industrial workers (hands above shoulders)	11.0
Frozen-food factory workers	9.4
Assembly line packers	8.1
Meat cutters	7.4
Shoe assembly workers	7.3
Packers	6.4
Data entry workers	4.9
Packaging and folding workers	3.9
Scissor makers	1.4

Source: Armstrong *et al.* (1993)

condition occurs in a sample of people drawn from the occupational group in question, to the frequency with which it occurs in a 'control' sample, who are not exposed to the same sort of risk.

One feature of Table 8.4 which is particularly worth noting is that data entry workers (the only group of keyboard users included) come fairly low down on the list. In other words, although data entry workers (and by inference other intensive keyboard users) are at an elevated risk level (compared with other people), the level of risk is by no means as high as it is for some of the industrial groups that have been studied. This runs contrary to the popular impression to 'RSI' being 'the keyboard users' disease'.

This was borne out in the great RSI 'epidemic' which swept Australia in the 1980s (see Figure 8.14). Even when the epidemic was at its height (in 1985/1986), the annual incidence of new cases remained highest in the blue-collar jobs which have been traditionally associated with injuries of this kind. The 'epidemic' itself, however, was caused to a very great extent by an increased reporting of such injuries in keyboard users. Because there are nowadays a very large number of keyboard users, the problem of keyboard injury must be treated as a very important one (see below).

The term 'RSI' is widely deprecated on the grounds that it is inexact and misleading. This is broadly the case. Discussions of the matter generally revolve around fairly arid issues of semantics which need not concern us here. The only point we need bother to note in this respect is that *repetition* as such is but one risk factor amongst many on the aetiology of these conditions – and it need not necessarily be a particularly important one in any individual case. The likelihood is that static muscle loading is of greater causative significance in the over-use injuries sustained by keyboard users; and likewise, in the repetitive, short-cycle time tasks of the industrial assembly line, force requirements and the nature and extent of the movements in question may be more important than repetition rates as such.

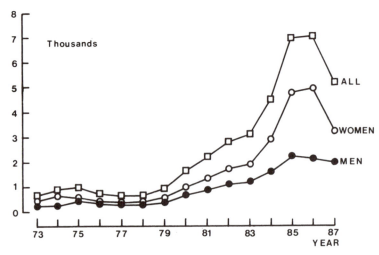

Figure 8.14   Annual incidence of repetitive strain injury (RSI) from New South Wales. (From S. Pheasant, *Ergonomics, Work and Health*, Macmillan, 1991, fig. 4.1, p. 79, reproduced with kind permission.)

In the UK, 'informed opinion' (as embodied, e.g., by the HSE) currently prefers the term 'work-related upper limb disorders', although 'repetitive strain injury' remains in more general use. In Australia where the term 'RSI' first originated (and likewise in New Zealand), it has been replaced by 'occupational over-use syndrome' (OOS). In North America, a broadly similar range of conditions are referred to as 'cumulative trauma disorders' (CTD); and in the Japanese literature the term 'occupational cervicobrachial disorders' (OCD) is used. It must be stressed that all of these terms are generic ones, which encompass a range of different 'clinical entities' (i.e. specific conditions).

8.5.1 On the varieties of RSI/WRULD

The chief difficulty with generic terms like 'RSI' and 'WRULD' is the diversity of clinical conditions to which they are applied. Two main subdivisions are generally recognized, which are sometimes referred to as 'Type I' and 'Type II' RSI respectively.

The former (Type I) are relatively discrete and localized over-use injuries to specific anatomical structures. This subdivision includes conditions resulting from traumatic inflammation of soft-tissue structures, such as the various forms of *peritendinitis* and *tenosynovitis* which affect the tendons of the muscles of the forearm and their soft-tissue coverings (and also muscles at other sites in the upper limb and shoulder region), as well as *lateral epicondylitis* and *medial epicondylitis* (otherwise known as 'tennis elbow' and 'golfer's elbow') which affect the points of origin of the extensor and flexor muscles respectively.

The Type I subdivision also includes the so-called *entrapment neuropathies*, although the term is not a particularly good one as the probability is that the symptoms of the condition may arise from irritation of the nerve as well as entrapment as such. The best known of these is *carpal tunnel syndrome* (which affects the median nerve as it passes through the confined space of the carpal tunnel at the front of the wrist). The median nerve may also be affected at other sites, as may the other nerves of the limb.

The underlying pathologies of the Type I conditions are relatively well understood (although some grey areas remain) and they are thus relatively uncontentious, except that in the medico-legal context there may be an entirely legitimate dispute as to whether the condition is caused by work in any individual case, as against being the consequence of 'normal wear and tear', 'degenerative changes', or a 'constitutional predisposition', etc. At law, the decision must be made in each individual case 'on the balance of probabilities', although for some conditions (e.g. peritendinitis and tenosynovitis) there is a stronger a priori assumption of work relatedness than for others (e.g. carpal tunnel syndrome), in that the available scientific evidence points to occupational or constitutional risk factors as being of greater or lesser relative importance in the (multifactorial) aetiology of the conditions in question.

It stands to reason that for almost any occupational or work-related condition one could name, some people will be more at risk than others – otherwise all the members of a particular work force would be affected rather than only some of them. This being so, it would seem to be fallacious to draw a distinction between a condition that is 'caused' by constitutional factors and merely 'aggravated' by work; as against being caused by work in a person who is at risk for constitutional reasons.

Over the last decade or so it has become increasingly clear, however, that very many people with RSI/WRULD cannot easily be allocated to any of the traditionally recognized clinical categories. These people – who are described as suffering from 'Type II' RSI – typically report symptoms of pain and dysfunction at multiple sites in the upper limb (or limbs), shoulder region and neck. These symptoms are often described as 'diffuse'. This is an unfortunate choice of word, in that it tends to imply that the symptoms are vague and insubstantial. They are not – at least, not always. In some cases they are crippling. A better description is 'disseminated'. It is likewise often said that these people have no objective clinical signs. (In medical parlance, a 'symptom' is something reported by the patient; a 'sign' is something that the physician observes for herself.) This is only partly true at best – in that the principal signs that may be observed are ones in which the patient reports pain – either on the palpation of tender structures (mainly muscles) or on the performance of certain diagnostic manoeuvres (the details of which need not concern us here).

The experienced examiner will also be able to detect palpable changes in the physical quality of the muscles, which may feel 'hard' or 'compacted', etc. In some cases there will be a change in the temperature of the affected limb, indicative of a disturbance of bloodflow.

The term 'RSI' (or alternatively 'repetitive strain syndrome' [RSS]) is sometimes applied to these disseminated conditions by default, and for want of a better alternative, as if it were a diagnosis. This use of the same term in both the generic and quasi-diagnostic senses has been a source of much avoidable confusion. The practice is to be deprecated. This leaves us, however, with the problem of what else to call these conditions. My own preference is for the term *disseminated over-use syndrome* (DOS).

Although, in this author's experience, the disseminated forms of RSI/WRULD are by no means unknown in manual workers on industrial assembly lines, they occur most prominently and characteristically in keyboard users. They are in fact the classic *keyboard injury* (see below). In contrast, assembly workers are most commonly affected by the localized varieties of RSI/WRULD. This suggests that different causative mechanisms are involved.

The disseminated over-use syndrome of the keyboard user has a characteristic natural history. The first symptoms are typically minor ones: most often a tingling in the hands or aching at the wrist; less often a dull ache in the neck or shoulder area. At first this comes on towards the end of the working day and subsides in the evenings and at weekends. The symptoms gradually become more severe, more unremitting and come to affect the person's activities away from the workplace, interfere with her sleep, and so on. And as they do so the symptoms 'spread': either proximally (up the limb) or distally (down the limb), as the case may be. In due course they may 'cross over' to the opposite limb; or they may come to affect the upper back, the breast, or the side of the face and even occasionally the low back and lower limbs.

The underlying pathology of the syndrome remains both obscure and contentious. There are those who take the view (often in the medico-legal context) that what is unknown is unreal. 'If I wasn't taught about this disease at medical school, and I don't know how to treat it, then how can it possibly exist?' They therefore say that people who claim to suffer from this syndrome are deluded; or they argue that the symptoms these people report are the product of 'conversion hysteria' or 'somatization' or other kinds of psychobabble. They are wrong. The syndrome

undoubtedly has a basis in organic pathology – and, complex as they are, the underlying mechanisms are slowly being unravelled.

In essence, the condition progresses from one of ordinary muscle fatigue which,
when opportunities for recovery are inadequate, becomes chronic. At some point
(and the mechanism is not well understood), a self-sustaining cycle of inflammation,
pain and muscle spasm supervenes, involving the activation in the muscle of what
are sometimes called 'trigger points' (see Wigley 1990, Pheasant 1991a, b etc.). At
the same time (in parallel as it were) a cascade of changes is initiated in the central
nervous system (CNS) mechanisms which mediate the experience of pain. This is
sometimes called *pain amplification* or *neurological sensitization* (Pheasant 1991a,b,

Table 8.5   A provisional clinical classification of work-related upper
limb disorders.

Type I	Localized condition, generally involving soft-tissue injury, e.g. tenosynovitis, peritendinitis, epicondylitis, carpal tunnel syndrome, and other entrapment neuropathies. Physical signs specific to the condition in question are present.
Type II	'Diffuse' condition involving pain and dysfunction at multiple sites. Tenderness on palpation in multiple muscle groups, often with signs of 'adverse neural tension' and sometimes with signs of neurovascular disturbance. Absence of physical signs of specific (i.e. Type I) conditions. The classic disseminated over-use syndrome (which occurs particularly in keyboard users).
Type III	Type I and II conditions are present concurrently.
Type IV	More than one Type I condition are present concurrently, e.g. carpal tunnel with epicondylitis.
Type V	Localized condition involving chronic pain and dysfunction. No evidence of any specific (Type I) condition, either now are at any prior stage in the history.
Type VI	Localized condition involving chronic pain and dysfunction with clear (?) evidence of a Type I condition at some prior stage in the history; but no ongoing physical signs thereof. Symptoms are very often activity related. May be regarded as a 'post-injury syndrome' resulting from a process of 'sensitization'.
Type VII	Disseminated (Type II) condition with clear (?) evidence of a Type I condition at some prior stage in the history; but no ongoing physical signs thereof. May be regarded as a 'post-injury syndrome'.

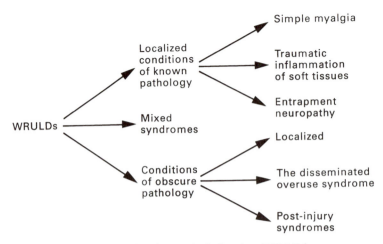

Figure 8.15   Classification of work-related upper limb disorders (WRULDs).

1992, 1994, Cohen *et al.* 1992, Gibson *et al.* 1991, Helme *et al.* 1992). Via various feedback loops in the CNS, disturbances of motor control and bloodflow may also ensue. The entrapment or irritation of peripheral nerves (which physiotherapists call 'adverse neural tension') may also be part of the picture, although how this relates to the other mechanisms is again not clear.

Psychological factors may well play a role in this process, but they are certainly not the sole causative factors involved – and there is good evidence that in many cases psychological symptoms that RSI victims report (anxiety, depression) are secondary or tertiary developments from the initial physical disorder (see below). In other words they are consequences, not causes. Sleep disturbance may be an intervening causative link.

The classification of the conditions falling into the overall RSI/WRULD category into Types I and II is a useful one – but it is something of an oversimplification. Some of the possibilities are summarized in Table 8.5 and Figure 8.15 (see also Pheasant 1994).

8.5.2   Over-use injuries to process workers

The over-use injuries to the hand, wrist and forearm, which are endemic in manual workers on industrial assembly lines, etc., are the product of a number of risk factors. The most prominent is very often a lack of task diversity, in that if you repeat the same wrist and hand movement, or short-cycle manipulative operation, endlessly throughout every working day, then it stands to reason that some anatomical structure or another in the hand or arm is simply going to wear out. This must surely be a matter of common sense. We could call this 'abnormal wear and tear' as against the 'normal wear and tear' of everyday life. A closely related risk factor is time pressure, whether this is imposed by the pace of the machine or by production targets, incentive schemes, etc., in self-paced work.

Over-use injuries are not solely restricted to short-cycle time tasks, which are repetitive in the narrow sense of the word. Some quite complex tasks may carry a high level of risk if the demands they impose are excessive in other ways. Likewise, although job rotation is always in theory to be encouraged, it may be of minimal benefit if the jobs between which the person rotates all have similar *biomechanical profiles* – that is, if they impose similar patterns of loading on the muscles and soft tissues.

The magnitude of the forces that it is necessary for the person to exert in the task in question is also relevant. In an elegant study of workers from a number of industrial plants, Silverstein *et al.* (1986) compared the prevalence of WRULDs in those whose jobs entailed high and low levels of force and repetitiveness respectively. Both force and repetition were statistically significant risk factors on their own – and the combination of the two carried a particularly high level of risk. This basic pattern of association was confirmed in a subsequent study of the specific disorder carpal tunnel syndrome, which was reported by the same authors, although in this latter study, force on its own was not a significant risk factor (Silverstein *et al.* 1987). The data of both studies suggest that when force and repetition are combined, their effects are at least multiplicative. Overall, these findings are just about what we should expect, on the assumption that the pathologies of these conditions are the consequence of 'abnormal wear and tear' which results from hand-intensive work.

It is generally recognized that the jobs that carry the highest levels of risk are ones in which forceful gripping actions are combined with turning actions and/or are made with a deviated wrist. In practice, turning actions will generally involve wrist deviation; but you can have a deviated wrist without making a turning action (e.g., when using certain badly designed hand tools).

Anatomical and biomechanical considerations would lead us to predict that there will be certain relatively consistent associations between particular disorders and particular patterns of movement. For example, we should expect 'clothes wringing' types of action (in which flexion and ulnar deviation are combined with supination of the forearm) to result in tenosynovitis or peritendinitis affecting the tendons of the extensors (particularly those that act on the thumb); and likewise that repetitive flexion and extension of the wrist would lead to carpal tunnel syndrome (either directly due to mechanical irritation of nerve, or secondary to a flexor tenosynovitis). (For an explanation of these anatomical terms, see section 5.2.)

Experience suggests that these predictable associations do indeed occur reasonably consistently – and such little epidemiology as there is would tend to confirm this (Pheasant 1991a). But they do not by any means occur infallibly – and in practice, just about any of the disorders in question may be associated with any of the motion patterns. In one way this is not particularly surprising. The functional anatomy of the hand is complex. All of the forearm muscles, whose tendons cross the wrist to insert on the bones of the hand, have multiple functions – as prime movers, synergists or muscles of stabilization. It would seem probable, furthermore, that the various anatomical structures of the 'muscle tendon unit' (i.e. the muscle itself and its soft-tissue attachments at either end) may be subject to over-use injury *both* when they repeatedly contract *and* when they are repeatedly stretched by their antagonists. So the number of potentially injurious permutations and combinations is considerable.

The particular type of gripping action that the task entails is also a factor. Pinch grips and claw-like grips (i.e. 'precision grips') both entail a higher internal biomechanical loading for a given externally applied force (and thus a higher level of risk) than full grasping actions (i.e. 'power grips'; see section 5.3). The 'overspreading' of the hand is also a risk factor.

There is epidemiological evidence that carpal tunnel syndrome may be caused by the use of vibrating tools (Cannon *et al.* 1981). It may also be caused by repeated impact as in using a hammer. Tenosynovitis of the finger flexors may likewise be caused by repeated blunt trauma – for example when using badly designed hand tools that have pressure 'hot-spots' on their handles.

It is widely recognized that people become accustomed to repetitive (and otherwise hand-intensive) work over a period of time. This process of adaptation is sometimes known as 'work hardening'. It probably has a number of physiological components. A simple muscle training effect is part of the story, as in all probability is the greater 'economy of movement' that comes with increasing skill. But the physiology may well be more complicated than this, and the process of adaptation may well be confounded with 'survivor effects' (i.e. those people who are unable to adapt leave the job).

Whatever the underlying physiology, however, it stands to reason that, given that such processes of adaptation do indeed occur, then newcomers to a particular job will be at an elevated level of risk as compared with 'old hands'. This has particularly been shown to be so for peritendinitis crepitans (Thompson *et al.* 1951). By the same token, we should expect any change in working practices that entails an increase in task demands to result in an increase in injury rate. Again this is borne out in practice. It also seems that people lose some of their physiological adaptation during holidays, periods of sickness absence and other lay-offs. (This is sometimes called de-adaptation.) So when they return to work they are at risk (Thompson *et al.* 1951).

It must be stressed, however, that these conditions are by no means to be regarded solely as 'training injuries', in that although unaccustomed work is clearly an important risk factor, the process of injury may also occur insidiously, leading to the onset of the condition in a seasoned worker and without any change in working practices being involved. It may be that when the condition arises in this way, it is because the person's capacity to tolerate hand-intensive work is diminishing with time (due to the cumulative effects of the work itself or to normal ageing) until the point is reached when the demands of the task come to exceed that capacity.

We are now in a position to summarize the principal ergonomic risk factors that are associated with over-use injuries to the hand, wrist and forearm in process workers. These are set out in Table 8.6.

Workers on assembly lines are also prone to over-use injuries to the muscles and soft tissues of the neck and shoulder region. These are most commonly caused by working for lengthy periods with the arms in a raised position, and/or from making frequent or repeated reaching actions (particularly overhead reaching or reaching behind the body).

Of the various WRULDs that affect the process worker, tenosynovitis has traditionally been regarded as the most important in the UK, whereas in the US carpal tunnel syndrome has the greater prominence, and in the Scandinavian countries ergonomic and epidemiological studies tend to focus on conditions affecting the neck and shoulder region (such as 'tension neck' or trapezius myalgia). It is difficult to account for these differences in emphasis. It seems wholly unlikely that they would reflect underlying differences in the true prevalence of these conditions. Nor does it seem likely that the conditions (at least as they are normally defined) can be mistaken for each other by anyone with more than a rudimentary knowledge of such matters.

Table 8.6 Over-use injuries to the forearm, wrist and hand in process workers – ergonomic risk factors.

- Lack of task diversity
- Time pressure
- Forceful exertion
- Frequent or repeated gripping actions, particularly if these are combined with turning actions and/or are made with a deviated wrist
- Pinch grips, claw grips, overspreading
- Vibration, impact, blunt trauma
- Unaccustomed work

8.5.3 Keyboard injuries

There is nothing at all new about keyboard injuries. The earliest reports of 'occupational cramps' in typists date back to not long after the invention of the typewriter (see Quintner 1991 for a historical review). In the first (1955) edition of his new classic *Diseases of the Occupations*, Hunter lists the occupational groups that are prone to suffer from tenosynovitis – and along with various groups of industrial manual workers and agricultural craft workers he includes on the list 'typists and comptometer operators'. (A comptometer was a primitive mechanical or electromechanical calculating machine, the keys of which sometimes had to be depressed in combinations like chords on a piano).

The rapid increase in the computerization of clerical and office work and introduction of the electronic keyboard, which occurred in the 1980s, led to a dramatic upswing in the reporting of these conditions. It is perhaps debatable whether the increase in reporting reflects a true increase in the underlying incidence of these conditions – we simply do not have the data to be certain either way – but assuming that it does (which on balance seems likely) then we must ask ourselves quite carefully why this should be the case.

The force required to depress the keys on a modern electronic keyboard is only about one-twentieth of that which is required on a mechanical typewriter; and the 'travel' of the keys is likewise very much less. There are also more subtle differences in keying action, in that the old-fashioned mechanical typewriter requires its user to modulate carefully the force with which each individual key was struck if the letters are to appear evenly on the page. There is no such requirement on an electromechanical or electronic keyboard.

Modern keyboards are very much thinner, furthermore, which in principle ought to reduce problems of anthropometric fit (see Section 6.1).

On the basis of these sorts of consideration we might perhaps expect the introduction of the electronic keyboard to result in a reduction in injury rate rather than otherwise. Writing as long ago as 1980, Cakir *et al.* predicted that the introduction of the VDU would result in a *reduction* in the incidence of *tenosynovitis* in typists; although interestingly enough they predicted an *increase* in the incidence of what they called 'shoulder–arm syndrome' – or what we might call the disseminated over-use syndrome or 'diffuse' RSI, etc. Broadly speaking, this seems to have turned out to be correct.

The changeover from mechanical to electromechanical typewriters of increasing degrees of complexity – and then from electromechanical typewriters to screen-based word processing systems – has led over the last few decades to a steady upward trend in typing speeds. Thus an average-to-good typist, who could reach 50 words per minute (wpm) on a mechanical machine, might reach 60 wpm on a basic electric typewriter and 70 wpm on a modern screen-based system – and nowadays, speeds of 90 wpm are not that unusual.

Common sense would perhaps tell us that the higher the keystroke rate (i.e. typing speed), the higher the risk. Overall there is little or no hard evidence that this is the case, although the absence of such evidence does not of itself rule out the possibility and there is little evidence to the contrary either. Physiologically you could argue it either way. It could, for example, be argued that what matters is not so much the actual number of key depressions as such (per minute, hour, etc.), as the proportion of that person's physiological capacity which that rate in question represents. This being so, if a person is required (for whatever reason) to work for lengthy periods at close to the limit of her individual capacity, she will be potentially at risk – especially if circumstances are disadvantageous in other respects (e.g., because of an unsatisfactory working posture). Thus a 50 wpm typist working flat out to meet a deadline would be exposed to much the same risk as a 90 wpm typist working under the same degree of pressure for the same length of time.

On balance it seems probable that the direct agent of injury in most cases of (disseminated) RSI/WRULD that we encounter in keyboard users is the static muscle loading which is contingent on a fixed working posture, rather than the repetitive motions of striking the keys. Electromyographic studies by Onishi *et al.* (1982) have shown that (in a badly designed workstation) this static loading may reach as much as 30% of the muscle's maximum capacity, which is more than enough to result in very significant degrees of local muscle fatigue if there are not relatively frequent pauses for rest and recovery (see section 6.6). Early electromyographic studies of typists (Lundervold 1958) showed that with the onset of fatigue, muscle activation spreads to groups that were initially quiet. So the scene is set for a more generalized muscle fatigue to set in.

There are those who take the view that the light action of the electronic keyboard leads the user to 'hold back', particularly if they have been brought up on machines that have a more positive feel. Journalists transferring from their old portables onto screen-based systems often say this. If it is so, then physiologically this would equate to an increase in static muscle loading. As a hypothesis it would lend itself to testing by electromyography – but the experiments have not been done. What is very clearly the case, however, is that with the increased computerization of office work there has been a progressive diminution in the task diversity of the keyboard user – and an increase in the extent to which her work involves the inputting of streams of data or text onto the machine *and nothing else* – not even the manual operation of the carriage return and the winding of paper onto the platen.

There is also good evidence for an 'exposure effect'. In a study that has not been quoted anywhere nearly as widely as it deserves, Maurice Oxenburgh (1984) compared the prevalence of 'RSI' in members of a particular organization who used the word processor keyboard for greater or lesser periods of time each day. The results are shown in Figure 8.16 The overall prevalence in this organization was 27%. But the prevalence varied from 9% in those who were on the machine for less than 3 h per day to 70% in those who used it for more than 6 h – which is equivalent to a

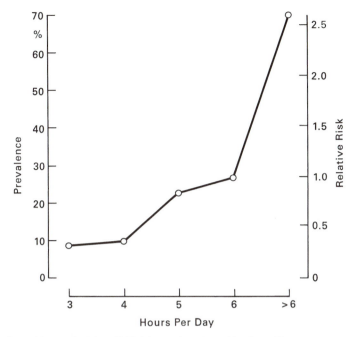

Figure 8.16   Repetitive strain injury (RSI) risk as a function of keyboard hours.

relative risk (compared with the overall prevalence) of 0.3 and 2.6 in the highest and lowest exposure categories respectively. The most striking feature of these data is the very sharp upswing in prevalence which occurs at past the 6 h point – which is sufficiently marked that we should be justified in regarding it a threshold exposure effect. The moral is simple. Nobody should use the word processor for more than 6 h per day.

To summarize, keyboard injury would appear to be associated with two main groups of risk factors in the working situation:

- Prolonged periods of intensive keyboard use in a constrained working posture, uninterrupted by rest pauses or changes of working activity, which may stem from lack of task diversity, pressure of work and so on.

- The overall degree of musculoskeletal loading inherent in the keying action, of which static loading rather than dynamic loading is probably the more important contributory factor. The most important determinant of the static loading that the task entails will in most cases be the person's working posture – which will often be contingent upon ergonomic deficiencies in the design or layout of the workstation.

Either or both classes of factors may be present in any particular case. Where both are present they may be presumed to act at least additively and more probably multiplicatively. Having said this, however, one does at times come across cases where neither of these risk factors seem to be present. This presents us with something of a difficulty. Two possibilities suggest themselves: either there are other occupational or environmental risk factors involved that we have not so far taken into account, or perhaps these people are exceptionally constitutionally vulnerable for reasons as yet unidentified.

Returning now to the matter of static loading, we may distinguish between *generalized* and *localized* sources of muscle tension in keyboard work. The electromyographic studies of Lundervold (1958) showed, for example, that working with a seat that is too high (relative to the floor) results in an increased tension in a number of muscle groups. We could regard this as a generalized muscle tension contingent on discomfort, postural instability, and so on. There may also be a more localized tension, for example in the forearm or shoulder girdle muscles as a result of working with a deviated wrist and abducted shoulder – because of a keyboard that is too high (relative to the seat), a desk top or keyboard that is too thick or a forward leaning position, etc. Or the user may rest her wrists on the edge of the desk (perhaps again because it is too high relative to the seat) and thus work with her wrists in extension, again leading to an increase in the static loading on the extensor muscles of the forearm. The studies of Duncan and Ferguson (1974) showed clear associations between postures of this kind and conditions that we would nowadays call RSI/WRULD. There was a particularly strong association between working with the wrists in extension or ulnar deviation and disorders affecting the muscles of the forearm. The need to rotate and incline the head and neck in order to read source documents placed flat on the table will likewise result in a static loading on the trapezius, sternomastoid and paraspinal neck muscles, leading to neck trouble. (For a further discussion of the ergonomics of keyboard workstations see Chapter 6).

Green and Briggs (1989) report an interesting study in which the anthropometric characteristics of keyboard users with and without symptoms of RSI/WRULD are compared. There was a tendency for those with symptoms to be more overweight, to have greater body breadths and lesser limb lengths. The authors discuss these findings in terms of lack of exercise and anthropometric mismatches with furniture and equipment. But it is worth noting that the combination of greater body breadth with lesser relative limb length will inevitably result in a greater degree of ulnar deviation of the keyboard – and thus a greater static loading on the forearm muscles.

In principle we should expect both generalized and localized muscle tension to be of causative significance in the aetiology of keyboard injury. Of the two, we should probably expect the localized tension to have the more decisive effect, but this by no means need be infallibly the case.

Psychological stress is a potent generator of (generalized) muscle tension. Could psychological stress be the decisive causative factor therefore in some cases of keyboard injury? This is a very difficult question indeed. The epidemiology certainly points to an association – at the statistical level, that is. In a study of data entry workers, Ryan and Bampton (1988) found that those who had symptoms in their necks, shoulders, arms and hands (and tenderness on palpation in relevant muscle groups) were more likely to report low levels of autonomy, peer group cohesion, and role clarity and to say that they found their work boring and stressful. (They also reported missing rest breaks more often.) These associations could of course be explained in a number of ways. The possibilities have been narrowed down somewhat by a subsequent study reported by Hopkins (1990) who compared *symptomfree* workers in organizations doing similar sorts of keyboard work but having high and low prevalences of RSI respectively. Those in the high prevalence workplaces

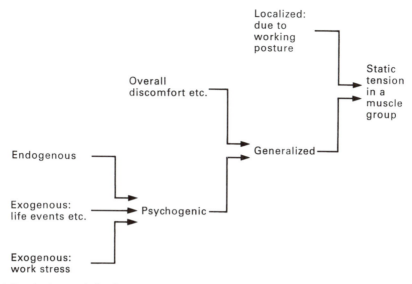

Figure 8.17 Static muscle loading in keyboard work: summation of contributory factors.

reported lower levels of autonomy, peer group cohesion, task diversity and job satisfaction, and higher levels of stress and boredom. Thus the same psychological
parameters that distinguish sufferers from non-sufferers (in Ryan and Bampton) also
distinguish non-sufferers in high- and low-risk working environments (in Hopkins).
By inference then, insomuch as psychological factors are of causative significance in
the aetiology of keyboard injury, this influence stems from the external psychosocial
features of the working situation rather than from the personal idiosyncrasies or
inner mental turmoils of the individual in question. The medico-legal consequences
of this finding are far-reaching.

To summarize: overall, the likelihood is that the proximate cause of keyboard
injury is static muscle tension. This tension can in principle stem from a number of
possible sources as shown in Figure 8.17.

The studies of Ryan and Bampton and Hopkins were, however, ones of people
reporting relatively minor symptoms rather than the fully developed syndrome. This
leaves the important question of whether such symptoms are more likely to progress
to severe and disabling problems in people who have 'vulnerable personalities'. An
elegant study reported by Spence (1990) indicates that this is not the case. She compared five samples of subjects. People suffering from RSI of recent onset did not
differ psychologically from people who had recently suffered other sorts of injuries
to their upper limbs. Neither did either of these groups differ from normal controls,
whereas people with longstanding RSI and people with chronic problems resulting
from injuries showed similar patterns of deviation from the psychological norm.
This shows quite clearly that the anxiety and depression that are so characteristic of
the RSI victim are the consequences of that person's physical condition, not its
causes. We have little reason to find this surprising.

PART TWO

The *Bodyspace* Tables

CHAPTER NINE

Human diversity

In this chapter we shall consider the principal ways in which *samples* and *populations* of human beings differ in their anthropometric characteristics – and the *biosocial* factors that underlie these differences.

The sizes, shapes and strengths of human beings are very often 'broken down by age and sex'. In defining a target population for anthropometric purposes, we must also take into account ethnicity, social class and occupation. Superimposed over these differences are changes occurring within populations over a period of time. Some of these are attributable to the migration and genetic admixture of hitherto distinct ethnic groups; others to more complex historical processes, which over the last century or so have led to an almost worldwide increase in stature, which is referred to as 'the secular trend'.

The extent to which these measurable differences between populations of human beings are determined by biological (or genetic) factors, as against social (or environmental) ones, poses a difficult set of equations. This 'nature/nurture' controversy has ramifications in very many branches of the human sciences. In reality, asking whether a given characteristic is determined by inheritance or by upbringing and lifestyle is probably a bit like asking whether the area of a field is determined by its length or its breadth.

One further point must be stressed. When comparing and contrasting the measurable characteristics of different groups of people, we will always be dealing with *within-group* variability as well as *between-group* variability. The greater the former compared with the latter, the less significant will be the difference between the groups (both in terms of statistical theory and in terms of ergonomics practice).

Consider humankind as a whole. It is debatable whether anthropometric data available at the present time, even if they could be assembled in one place, would constitute a representative sample of all human beings living at the present time. Such indications as we have, however (see Tildesley 1950), suggest that, in round figures, the stature of all adult living male adults has a mean value of about 1650 mm with a standard deviation of 80 mm – which, the reader will recall from Chapter 2, we shall write as 1650 [80] mm. Assuming a sex difference in average stature of

7% and an equal coefficient of variation, the stature of living female adults will have a distribution of about 1535 [75] mm. (These figures could doubtless be improved by anyone with the patience to do so, but will serve as a starting point.)

The adult population of Great Britain is well into the taller half of the human race: 1740 [70] mm for men and 1610 [62] mm for women. Hence the average British adult male is about 87th %ile for the human race as a whole.

According to Roberts (1975) the shortest people in the world are the Efe and Basua 'pygmies' of central Africa, whose average stature is 1438 [70] mm for men and 1372 [78] mm for women. The tallest are the Dinka Nilotes of the southern Sudan: 1829 [61] mm for men and 1689 [58] mm for women. However, differences almost as great as these may be found between particular samples drawn from the British population. Guardsmen (Gooderson and Beebee 1977) stand at some 1803 [63] mm whereas a sample of elderly women measured by Caroline Harris and her colleagues (Institute for Consumer Ergonomics 1983) had a stature of 1515 [70] mm (deducting a modest 20 mm for shoes since the subjects were measured shod).

The human race is more varied still. The limits of what is normally considered to be 'clinical normality' are set at an adult stature of something in the order of 1370 and 2010 mm. (The exact figures quoted vary somewhat, being essentially arbitrary.) According to the best information available at the time of writing (as set out in the *Guinness Book of Records*) the shortest living adult is 570 mm tall; and the tallest is 2015 mm.

9.1 Sex differences

It has become fashionable of late to refer to the differences between men and women as ones of 'gender' rather than 'sex'. This is incorrect. The word 'gender' applies to the distinction that exists in most European languages (other than English) between 'masculine' and 'feminine' nouns, rather than the differences between male and female living organisms. The human species, like all the higher animals, is sexually dimorphous.

Are the anthropometric differences between men and women attributable to underlying biological (i.e. genetic and physiological) differences, or to cultural differences in upbringing and lifestyle? We can be fairly sure that sex differences in stature and related body dimensions and most differences in bodily proportions are almost entirely biological in their origin, although there may be a small overlay of differences attributable to lifestyle, etc. In the case of muscular strength, however, the position is more evenly balanced, and although the male of the species has the greater physiological propensity to the acquisition of muscle strength, the overlay of differences associated with physical training and lifestyle is very considerable.

What is the best way of describing sex differences statistically? The most frequent found in the literature is a straightforward comparison of means. Hence, we read statements like 'on average women are 7% shorter than men' or 'on average women are 65% as strong as men'. Let us call the average female dimension (or strength) divided by the average male dimension (or strength) the F/M ratio for short. However, for all the variables we are likely to consider in the present text, there is considerable overlap between the male and female distributions. The F/M ratio of means tells us very little about this combined distribution. (Among many other equally interesting descriptions we might include the ratio of the 95th %ile female to the 5th %ile male or the 5th %ile female to the 95th %ile male; the percentage of women stronger than the 5th %ile man or men weaker than the 95th %ile woman, etc.).

At the very least, a descriptive index should reflect both the difference between the means and the magnitude of the variances of the male and female distributions under consideration. It would be useful and informative to know the proportion in the total variance in strength (i.e. in the combined unisex distribution) which is attributable to strength. Afficionados of the one-way analysis of variances will understand that this index is given by the equation:

$$R^2 = \text{between sex Ssq/total Ssq} \tag{9.1}$$

(If this equation is absolute gibberish to you, don't worry too much – alternatively turn to any textbook of statistics.)

When preparing a paper on sex differences in strength a few years ago (Pheasant 1983), I paused to ask myself just what does the layperson have in mind when he (or she) asks 'How true is it that men are stronger than women?' Consider a population of men and a population of women. Suppose we select a man at random followed by a woman at random and compare their strengths. We will call such a comparison a chance encounter. If we perform an infinite number of such comparisons we may generate a statistical distribution of chance encounters. The F/M ratio is equivalent to an encounter between an average woman and an average man. Both the layperson and the human scientist wish to know about the remainder of the distribution. For reasons that would only be comprehensible to a competent mathematical statistician, the distribution of the ratios of two normal distributions is not itself normally distributed. If, however, we forget about ratios and consider absolute differences the problem becomes much more tractable. If differences are used, the distribution of chance encounters is normal and its parameters are given by:

$$M_e = M_m - M_f \tag{9.2}$$

$$S_e^2 = S_m^2 + S_f^2 \tag{9.3}$$

where M and S are the mean and standard deviation; the subscripts m and f refer to men and women, respectively. The value of zero in this distribution represents a chance encounter between a man and a woman of equal strength. It is simple to calculate the proportions of the distribution lying on either side of this point (by calculating z and looking up a table as described in Chapter 2). We therefore know the percentage of chance encounters in which the female is stronger or, in the more general case, the percentage of chance encounters in which the female exceeds the male (% CEFEM). This index is as close as we can reasonably get to the layperson's conception of the question.

Obviously enough, any investigation of sex differences will founder if the samples of men and women who are studied are not truly comparable. Thus a comparison of male navvies with female secretaries or of male secretaries with female navvies, is not solely an investigation of sex differences *per se*.

In general men will exceed women in all linear body dimensions except hip breadth (as shown in the data in the tables that follow). There are ethnic differences in the magnitude of these sex differences – at least for stature. Eveleth (1975) found greater differences in Amerindians than in Europeans, who in turn showed greater differences than black Africans.

Many sex differences in body proportion are too well known to require further comment. In general the lengths of the upper and lower limbs are proportionally as well as absolutely greater in men. Thus the ratio of sitting height to stature (sometimes called 'sitting height index' and used as an index of relative trunk length) will be greater in women than in men. The only limb dimension that is proportionally greater in women is buttock–knee length – this being due to differences in the form of the male and female buttock. There is no difference between men and women in the proportional values of either head length or head breadth.

In addition to the dimensional anthropometrics described above, men and women differ in their bodily composition. In general, fat represents a greater proportion of body weight in the adult female than in the male. (Subcutaneous fat is also distributed differently, women having a propensity to accumulate fat in the breasts, hips, thighs and upper arms. Abdominal fat accumulates above the umbilicus in men and below the umbilicus in women.) The most direct way of measuring body fat is by densitometry. Fat is a good deal less dense than lean tissue, so if the density of the body is determined (usually by underwater weighing) it is possible to calculate the percentage that fat contributes to the weight of the body. Durnin and Rahaman (1967) found this percentage to be 13.5 [5.8] for adult men and 24.2 [6.5] for adult women (F/M = 179%, R^2 = 43%; % CEFEM = 89).

I have previously published a detailed analysis of sex differences in strength (Pheasant 1983). A survey of the literature located a total of 112 datasets in which a direct and, presumably, valid comparison of the performances of men and women in some test of static strength could be compared. Indices of sex differences were calculated for each of these datasets (see Table 9.1). Although the average value of the F/M ratio is 61% – very close to commonly quoted figures of women being two-thirds as strong as men – the ratios found in the whole series range from 37 to 90%. The other indices tell a similar story – sex can account for a major (85%) or a negligible (3%) proportion of the total variance in strength.

An interesting pattern emerges if we divide the datasets into groups according to the part of the body tested. Upper limb tests show greater sex differences than lower limb tests, or tests of pushing, pulling and lifting actions, with tests of trunk strength being somewhere between the two. Subdivision of the upper limb category revealed

Table 9.1 Sex differences in strength (classified by part of body, etc., tested).

Part of body	No. of tests	F/M (%)			R^2 (%)			% CEFEM		
		Min.	Ave.	Max.	Min.	Ave.	Max.	Min.	Ave.	Max.
Lower limb	17	50	66	81	11	43	62	4	12	28
Push/pull/lift	41	38	65	90	3	33	72	1	14	37
Trunk	11	37	62	68	35	54	61	1	6	13
Upper limb	29	44	58	86	7	53	77	1	8	35
Miscellaneous	14	43	53	61	35	63	85	0	5	16
All tests	112	37	61	90	3	45	85	0	10	37

Source: Data from Pheasant (1983).

that tests of hand and forearm muscles gave lesser sex differences than tests of upper arm and shoulder muscles. Taking all three indices together, the least sex differences were found in the push/pull/lift category. (The factors that determine performance in a task of this kind are numerous and complex. The co-ordinated activity of many muscles may be involved and in some cases the limiting factor may be body weight and its leverage.) Hettinger (1961), considering the magnitude of the F/M ratio in various muscle groups, suggested that the sex difference is small in those muscle groups that are under-used in everyday life. In the present author's view it would be just as reasonable to propose the opposite hypothesis on the basis of the available data – and the opposite would be just as biologically plausible!

What is the underlying physiology of sex differences in strength? The strength of a muscle is directly proportional to the effective cross-sectional area of its contractile tissue. The cross-sectional area of a muscle must be closely related to the bulk that is visible to the casual observer, or can be measured with a tape. Ikai and Fukanaga (1968), using a sophisticated ultrasound measurement of cross-sectional area, found strength of approximately 6.5 kgf/cm^2 of muscle tissue which was independent both of sex and of age from 12 years upwards. Trained judo men had the same strength per unit area as untrained people. In general then, it is the quantity not the quality of muscle that counts; at least in strength measurements of short duration.

It is widely accepted that the 'secondary sex differences' of fat distribution and muscle bulk result from the relative concentrations of the sex hormones – androgens in the male and oestrogens and progestogens in the female. Testosterone, the most important of the androgens, is produced in large quantities in the testis, but also in very small quantities in the ovaries. (The level of testosterone in the blood plasma of men is 20–30 times that of women.) In response to a given training programme, men show a faster and greater increase in strength than women; and this difference is generally attributed to the effects of testosterone (Hettinger 1961, Klafs and Lyon 1978). Brown and Wilmore (1974) monitored a small group of female throwing-event athletes over a six months' maximal resistance training programme. Strength increased by 15–53% but there was little evidence of muscular hypertrophy (increase in bulk). How this latter finding is to be reconciled with the results of Ikai and Fakanaga (1968) is not yet clear.

Klafs and Lyon (1978) speculate that women who have a relatively high level of plasma testosterone will 'bulk up' like men in response to intense weight training and, indeed, the current vogue for female body building shows that a striking degree of muscular hypertrophy may occur in some individuals. It has often been suggested that female athletes are in one respect or another less 'feminine' than their more sedentary sisters. Malina and Zavaleta (1976) calculated an 'androgyny score' of 3 biacromial (shoulder)–bicristal (pelvic) breadth. (Hence a high score on this index is indicative of a masculine skeletal frame and a low score is feminine.) Runners (both long and short distance) did not differ from non-athletes in androgyny scores but jumpers and throwers were significantly 'masculine' according to this criterion. Is this a training effect or does it represent self-selection? The latter is generally believed to be the case amongst physical educationalists (Klafs and Lyon 1978). Adams (1961) compared young black women who had been engaged in heavy farm labour all their lives with ones who had not. Although the labourers were larger in óverall size and muscular development than the controls, the 3 biacromial–bicristal androgyny score was similar for the two groups.

Our discussion of these matters would not be quite complete without some passing reference to 'norms', 'ideals', 'cultural expectations' and the elusive phenomena of taste and preference. The history of European art reveals considerable diversity in the ideal female form – consider, for example, the way that Venus was depicted by Rubens, Titian, Botticelli and Cranach, to name but four in order of decreasing radius of curvature. The ideal male form (Mars, Adam, etc.) has remained remarkably constant by comparison (or is it just that I don't notice the differences?). The one empirical study of these matters that I have discovered is that of Garner *et al.* (1980) who took the strikingly original approach of analysing the recorded heights, weights and bodily circumferences of all *Playboy* magazine centrefolds between 1959 and 1978. The trend over the period was for an increase in height, reduction in weight for height, bust circumference, and hip circumference and increase in waist circumference – indicating a move towards a body form that the authors characterized (somewhat oddly) as 'tubular'. I have not had the opportunity of determining to what extent this trend has continued. There must presumably be a limit.

9.2   Ethnic differences

An ethnic group is a population of individuals who inhabit a specified geographical distribution and who have certain physical characteristics in common which serve, in statistical terms, to distinguish them from other such groups of people. These characteristics may be presumed to be predominantly hereditary, although the extent to which this is the case is sometimes contentious.

Ethnic groups may or may not be co-extensive with national, linguistic or other boundaries – hence the various ethnic types to be found within the population of Europe are distributed across national boundaries, although the frequency with which a given type is encountered will vary from place to place. To some extent ethnic groups fall into more or less natural clusters, which may be referred to as the Negroid, Caucasoid and Mongoloid 'divisions' or 'major groups' of humankind. The term 'race' has tended to disappear from the scientific literature, due, one might suppose, to a collective embarrassment occasioned by its misuse for dogmatic and propagandist purposes. As Gould (1984) has clearly shown, supposedly objective scientific writers have colluded in this misuse.

The Negroid division includes most of the dark-skinned peoples of Africa, together with certain minor ethnic groups of Asia and the Pacific islands. The Caucasoid division includes both light- and dark-skinned peoples resident in Europe, North Africa, Asia Minor, the Middle East, India and Polynesia (together with the indigenous population of Australia and some other ethnic groups who form a subdivision of their own). The Mongoloid division comprises a large number of ethnic groups distributed across central, eastern and south-eastern Asia, together with the indigenous populations of the Americas.

Samples of adults may vary from each other either in overall size (as measured by stature or weight) or in bodily proportions. The most characteristic ethnic differences are of the latter kind since the major divisions of humankind include both tall and short populations. Figure 9.1 illustrates some salient features. Average sitting height (measured from the seat surface) has been plotted against average stature. The ratio of the two (relative sitting height) is plotted as oblique lines on the chart.

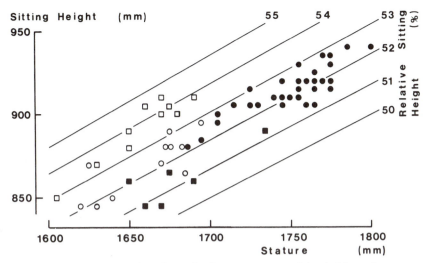

Figure 9.1 Ethnic differences in the relationship between average sitting height and average stature in samples of adult men. ● = European (including samples of predominantly European descent); ○ = Indo-Mediterranean; □ = Far Eastern; ■ = African.

When relative sitting height is large, the sample is 'short legged' and vice versa. The data points are all male samples taken from Eveleth and Tanner (1976) and NASA (1978). Samples drawn from the civilian or military populations of the US (of which there are a considerable number in the literature) are classed as 'of predominantly European descent' – notwithstanding the fact that around 10% of the membership of such samples are of identifiably different ethnic origins.

Black Africans have proportionally longer lower limbs than Europeans; Far Eastern samples have proportionally shorter lower limbs, the difference being most marked in the Japanese, less in the Chinese and Koreans and least in the Thai and Vietnamese. These differences of proportion occur throughout the stature range. If we consider the European data only, there is a tendency for the ratio of sitting height to stature to be slightly greater for short samples than tall ones – suggesting the interesting hypothesis that the lower limbs contribute more to differences in stature than the trunk. The populations of Turkey, the Middle East and India, labelled 'Indo-Mediterranean', have proportions similar to Europeans but, typically, a lesser overall stature.

Do these ethnic differences in size and proportion have any evolutionary significance? Zoologists have identified two rules concerning morphological variations of warm-blooded polytypic species, of which humankind is an example. Bergman's rule states that the body size of varieties increases with decreasing mean temperature of the habitat. Allen's rule states that the relative size of exposed portions of the body decreases with decreasing temperature. Roberts (1973), in an extensive survey of the anthropometric literature concerning the world's indigenous populations, showed that these rules are in general applicable to humankind. Body weight is negatively correlated with mean annual temperature. Samples with the lowest body weights are not found outside the tropics, and the highest body weights are not found at latitudes lower than 30°. Furthermore, linearity of bodily form (as indicated by high values for relative limb lengths) shows a strong positive correlation with mean annual temperature. Taken together these findings indicate that ethnic groups inhabiting hot climates will tend to have a high ratio of surface area to body mass – which is advantageous for the loss of heat. Similarly, the inhabitants of cold regions are adapted for heat retention. Roberts concluded, however, that there were differences in form between the major ethnic divisions of humanity even when the effects of temperature have been taken into account.

The relative lengths of the upper limbs show a similar pattern of ethnic differences to the lower limbs, and there is some evidence to suggest that the differences are more due to a lengthening or shortening of the distal segment of the limb (i.e. the forearm or shank) than the proximal segment (i.e. the upper arm or thigh). The shoulders are a little narrower relative to stature in Africans than Europeans and the hips are considerably narrower in both sexes. In general, African bodily proportions are best described as 'linear'.

It would be a mistake to consider these differences in bodily size or shape to be fixed and immutable characteristics of ethnic groups. Several studies of migrant samples have shown significant differences between the growth patterns or adult dimensions of individuals born in the new environment and equivalent samples in the 'old country'. Boas (1912) and Shapiro (1939) are classic studies of this kind and subsequent investigations include Kaplan (1954), Greulich (1957) and, more recently, Koblianski and Arensburg (1977). Shapiro (1939) studied Japanese immigrants to Hawaii. He showed that although the Hawaiian-born generation are taller than the immigrants, and larger in most other dimensions, the ratios of the major bodily dimensions (i.e. relative sitting height, relative biacromial breadth) are not very different. This relative constancy of proportion has been confirmed by Miller (1961). This led Roberts (1975) to conclude that 'the data suggest a strong genetic component to body proportion, and a more labile overall size'.

Things are not quite this simple, however. There is evidence, for example, that the Japanese are becoming more like Europeans in terms of their relative limb lengths (Tanner *et al.* 1982, see below), but less like Europeans in terms of their head shapes (Yanagisawa and Kondo 1973). This seems very curious indeed.

It is generally believed that the anthropometric differences between European (or North American) user populations are sufficiently great that a product or item of equipment designed for the former will be unsuitable for the latter. The same may well be true of some Third World populations.

In the case of ethnic minorities in a working population of predominantly European descent, the situation is less clear; but given the relative magnitudes of the within-group and between-group variation concerned, we should not in general expect their presence to begin to be significant in ergonomic terms unless they made up more than about one-third of the total. This is only a rough rule of thumb, however, and there may be circumstances in which the presence of ethnic minorities in a working population is more critical. Thompson and Booth (1982), for example, suggest that there are circumstances in which people from certain ethnic groups may be more at risk if industrial safety guarding standards are not modified to take into account their particular anthropometric characteristics.

9.3   Growth and development

At birth we weigh some 3.3 [0.4] kg, and we are 500 [20] mm in length, of which our trunks represent some 70%. In the two decades that follow, our body length increases between three- and fourfold, our weight increases around 20-fold and our linear proportions change so that in the adult state the length of the trunk accounts for only 52% of the stature. However, the adult condition is by no means stationary – our bodily proportions are modified by our lifestyles and the inevitable processes of ageing. The anthropometrist who wishes to chart this course (or part of it) may most conveniently do so by a cross-sectional study in which several samples of individuals, representative of different age bands, are measured at the same time. (A cross-sectional age-band sample is known as a 'cohort'.) Data gathered by this means have certain limitations. In the case of children, only a very crude estimate can be obtained of the rate at which changes are taking place. Furthermore, our differences may be confounded with the effects of a secular trend. To disentangle these effects it is necessary to conduct longitudinal studies in which a sample of people are followed over an extended period of time.

The genetic and environmental factors that control human growth have been documented in detail by Tanner (1962, 1978), who has also published standards for the height and weight of British children which have been widely adopted in medical practice (Tanner *et al.* 1966, Tanner and Whitehouse 1976). The pattern of growth of a 'typical' boy and girl based on these data is shown in Figure 9.2 (The 'typical' child is a purely fictitious individual who is average in all respects at all ages.) At ages up to 2 years measurements are made on a supine infant; subsequently in a standing position. The rate of growth in boys is very rapid during infancy, declining steadily to reach its minimum at $11\frac{1}{2}$ years; it then accelerates again to reach its peak at 14 years before steadily decelerating as maturity is approached. The velocity peak around 14 years, known as the 'adolescent growth spurt', is associated with the events of puberty. The peak in the chart is broader and lower than it would be for any actual individual child since it represents the average of a sample of boys, all of whom are accelerating at different times. Hence, at 14 years some boys will have almost completed their growth spurt whereas others will scarcely have commenced it. As a consequence the standard deviations of the bodily dimensions of samples of

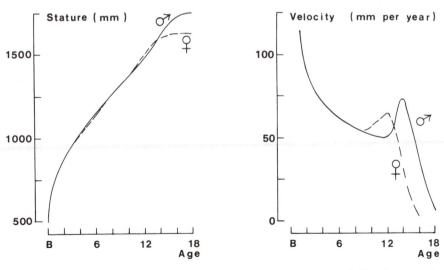

Figure 9.2 Growth from birth to maturity of a typical boy and girl: stature (left) and velocity (i.e. rate) of growth in stature (right). (Data from Tanner *et al.* (1966).)

adolescents are very large (see the tables in Chapter 10). The typical girl is a little shorter than the typical boy from birth to puberty but the growth spurt commences earlier in girls – at around 9 years reaching its maximum velocity at around 12 years, growth being more or less complete by 16 years. Hence, there is a period from about 11 to $13\frac{1}{2}$ years when the typical girl is taller than the typical boy. The typical boy reaches half his adult stature a few months after his second birthday, and the typical girl a few months before, although these figures will, of course, be subject to considerable variations in the population as a whole.

In addition to increasing in size, the human body changes considerably in shape. If the shape and composition of the body were the same throughout we would expect body weight to grow with the cube of stature (since weight is directly proportional to volume, assuming constant density). That would give an individual of average birth size, and who achieved an average male adult stature of 1740 mm, a body weight of 139 kg – which is close to twice the correct figure. In reality, growth is accompanied by an attenuation of bodily proportions.

Tanner (1962) has pointed out that there are various 'maturity gradients' which are superimposed upon the growth curve of the body as a whole – hence, at any point in time the upper parts of the body (particularly the head) are closer to their adult size than the lower parts; the upper limbs are further developed than the lower; and the distal segments of the limbs (hands, feet) are ahead of the proximal (thighs, arms). Cameron et al. (1982) also showed differences in the timing of the adolescent growth spurt for different parts of the body.

It is generally assumed that these gradients operate in such a way as to give a steady unidirectional transition from the large-headed, short-legged form of the child to the typical proportions of the adult. Tanner (1962) copied an illustration of this from Medawar (1944), who in turn took his from an anatomy textbook of 1915, which in turn is based on nineteenth-century data.

Medawar (1944) made the following statement: 'Just as the size of the human being increases with age, so, in an analogous but as yet unformulated way, does his shape. The property is best expressed by saying that change of shape keeps a certain definite trend, direction or 'sense' in time; like size, it does not retrace its steps.' Numerous authors have fitted mathematical equations to these supposedly simple transformations and some have attached biological significance to the constants in the equations.

Some years ago, whilst compiling anthropometric estimates for British schoolchildren, I chanced on certain discrepancies which lead me to believe that the assumption of a simple unidirectional change in shape was incorrect. Figure 9.3, previously published in Pheasant (1984a), is based upon the cross-sectional study of the under-18-year-old population of the US published by Snyder et al. (1977). The mean value of each dimension for each age cohort has been divided by the mean value of stature (or supine crown–heel length for the under-2-year-olds). In some dimensions, such as head length, we may observe the smooth unidirectional approach towards adult proportions which we have been led to expect – but these are the exceptions rather than the rule. Most dimensions show what might be termed a 'developmental overshoot'. Sitting height, for example, has achieved its adult percentage of stature by 9 years in girls and 11 years in boys; it then overshoots and reaches a minimum at the time when the adolescent growth spurt is at its peak (12 years in girls, 14 years in boys), before climbing back to its adult proportions. Knee height, as one might expect, shows a pattern that is similar but

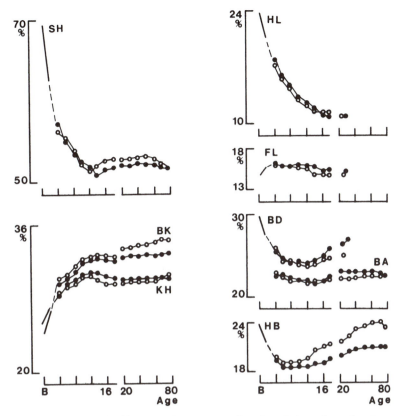

Figure 9.3 Effects of age on bodily proportions expressed as the relative values of various dimensions (% stature). SH = sitting height; BK = buttock–knee length; KH = knee height; HL = head length; FL = foot length; BD = bideltoid breadth; BA = biacromial breadth; HB = hip breadth; ● = boys and men; ○ = girls and women. (Original data from Synder *et al.* (1977), Stoudt *et al.* (1965, 1970) and NASA (1978).)

inverted, as to a lesser degree do shoulder–elbow and elbow–fingertip lengths (not shown in the figure). Both shoulder and hip breadths are proportionally large in early infancy and pass through proportional minimum – during adolescence in the former case and childhood in the latter. Foot length has a long plateau of elevated proportions in childhood before commencing a descent during adolescence. In summary, the data confirm the popular stereotypes of the 'dumpy' infant and the 'gangling' adolescent.

The data of Figure 9.3 are also interesting with respect to sex differences and the ages at which the bodily proportions of boys and girls first diverge. In the case of sitting height, knee height and foot length the divergence is associated with the events of puberty and the developmental overshoot. The bony pelvis of the female is broader than that of the male at birth (Tanner 1978) and there is a slight sex differ-ence in proportional hip breadth at the youngest age for which we have data – hip breadth also shows a slight divergence at around 6 years and a pronounced one at adolescence, which continues well into adulthood. (Buttock–knee length is quite similar so we are certainly dealing with soft-tissue upholstery to a large extent.) By

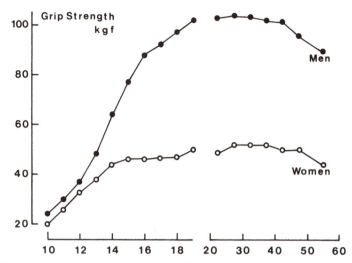

Figure 9.4 Effects of age and sex on grip strength. (Data from Montoye and Lamphier (1977).)

contrast, shoulder breadth (bideltoid, biacromial) does not show any measurable divergence until as late as 17 years.

The muscular strengths of boys and girls are similar during childhood and diverge at around the time of puberty, as shown in Figure 9.4 which is based on the data of Montoye and Lamphier (1977).

The age at which we reach 'anthropometric adulthood' is by no means as clear-cut as one might suppose. Growth standards conventionally stop at 16 years for girls and 18 years for boys. The growth of a long bone occurs by cell division in plates of cartilage which separate the ends (epiphyses) from the shaft (diaphysis) – when this cartilage finally turns into bone, growth cases (eiphyseal fusion). The clavicles continue to grow well into the twenties and so, to a lesser extent, do the bones of the spine. Andersson *et al.* (1965) demonstrated an increase in sitting height in a majority of boys after 18 years and girls after 17 years and in some boys after 20 years. A sample of Americans studied by Roche and Davila (1972) reached their adult stature at a median age of 21.2 years for boys and 17.3 years for girls; but some 10% of boys grew after 23.5 years and 10% of girls after 21.1 years. According to Roche and Davila (1972) this was partially due to late epiphyseal fusion in the lower limbs and partially to lengthening of the spine. Miall *et al.* (1967), in a longitudinal study of two Welsh communities, found evidence that men might grow slightly in stature well into their thirties.

9.4   The secular trend

Human biologists use the term 'secular trend' to describe alterations in the measurable characteristics of a population of human beings occurring over a period of time. Over a period of at least a century biosocial changes have been occurring in the population of much of the world which have led to:

- increase in the rate of growth of children;

- earlier onset of puberty, as indicated by menarche (the onset of the menstrual cycle) in girls and the adolescent growth spurt in both boys and girls;

- increase in adult stature, with a possible decrease in the age at which adult stature is reached.

The extensive statistical evidence concerning these changes has been reviewed by, amongst others, Tanner (1962, 1978), Meredith (1976) and Roche (1979).

Tanner (1962, 1978) summarizes the available evidence and concludes that from around 1880 to at least 1960, in virtually all European countries (including Sweden, Finland, Norway, France, Great Britain, Italy, Germany, Czechoslovakia, Poland, Hungary, the Soviet Union, Holland, Belgium, Switzerland and Austria), together with the US, Canada and Australia, the magnitude of the trend has been similar. The rate of change has been approximately:

- 15 mm per decade in stature and 0.5 kg per decade in weight at 5–7 years of age;

- 25 mm and 2 kg per decade during the time of adolescence;

- 10 mm per decade in adult stature.

This has been accompanied by a downward trend of 0.3 years per decade in the age of menarche. Roche (1979) points out that secular changes in size at birth have been small or non-existent.

Although the magnitude of changes in Europe and North America have been fairly uniform they are by no means universal. Japan, for example, has shown a particularly dramatic secular trend. The data of Tanner *et al.* (1982) show that in the decade between 1957 and 1967 Japanese boys increased in stature by:

- 31 mm at 6 years;

- 62 mm at 14 years;

- 33 mm at 17 years.

In the 1967–1977 period, however, these figures had declined to:

- 17 mm at 6 years;

- 35 mm at 14 years;

- 19 mm at 17 years.

This suggests that the explosive biosocial forces driving the change are beginning to wear themselves out. In contrast, Roche (1979) cites evidence that in India, and elsewhere in the Third World, there has actually been a secular decrease in adult stature.

If people are increasing in size, are they also changing in shape? The remarkable Japanese secular trend seems to be associated with an increase in the relative length of the leg – as the data of Tanner *et al.* (1982), plotted in Figure 9.5 show. It is doubtful, however, whether such a change of proportion is general. Figure 9.6 shows the relative sitting heights of samples of young American males (average ages between 18 and 30 years) plotted against the year in which the measurement was taken. There is no evidence of a secular trend in adult proportions. (This conclusion has been confirmed by Borkan *et al.* 1983.)

It is interesting to speculate as to whether our distant forebears were as short as we might imagine from recent secular trends. Anecdotal evidence concerning a range of artefacts from doorways to suits of armour abounds. Although it is not

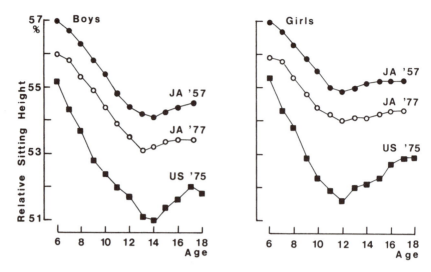

Figure 9.5   Secular trend in the bodily proportions of Japanese children (JA) compared with those of US children. Original data from Tanner *et al.* (1982) and Snyder *et al.* (1977).

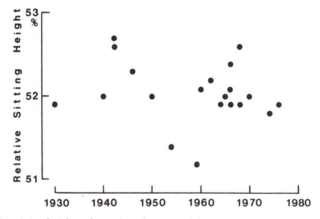

Figure 9.6   Relative sitting heights of samples of young adult US men measured between the years 1930 and 1980. Note the absence of any detectable secular trend in bodily proportions.

possible to calculate stature accurately from poorly preserved skeletal remains, the long bones of ancient burials allow us to make a reasonable estimate. The archaeological evidence summarized by Wells (1963) suggests that the statures of British males from neolithic to medieval times have always fallen within the taller part of the present-day human race. Indeed, figures quoted include average heights of 1732 mm for Anglo-Saxons and 1764 mm for Round Barrow burials, the latter actually exceeding the average height of present-day young men. The secular trend then seems to be a recovery from a setback which occurred somewhere in post-medieval times. Tanner (1978) cites various evidence that in the earlier part of the nineteenth century trends were small or absent, and plausibly associates them with the Industrial Revolution.

What then are the determining factors that have led to the phenomenon of the secular change? Speculation has been intense on this subject – most writers maintaining a cautious tone in their conclusions. Social/environmental influences such as the improved nutritional quality of diet and the reduction of infectious disease by improved hygiene and health care are the factors that most readily spring to mind, to which we might add the effects of urbanization and reduced family size, but we cannot ignore the possible influences of genetic factors such as heterosis, the beneficial effects that are said to derive from outbreeding and the breaking up of genetic isolates. A century ago most people married and raised their children within the confines of isolated communities; today we are approaching the condition of the 'global village'. As Tanner (1962) perspicaciously observed, 'it has been shown in several West European countries that outbreeding has in fact increased at a fairly steady rate since the introduction of the bicycle'.

The consensus view amongst human biologists tends to favour the environmental rather than the genetic causes. It seems most likely that genetic endowment sets a ceiling level to an individual's potential for growth and that environmental circumstances determine whether this ceiling is actually reached. If this is indeed the case, the end of the secular trend is in sight, at least in the economically developed countries of Europe, North America and elsewhere – since we could reasonably argue that the further amelioration of environmental conditions, beyond those adequate for the achievement of full genetic potential, cannot lead to any further changes. Considerable evidence suggests that this limit has indeed been reached, at least in some communities. Backwin and McLaughlin (1964) showed that Harvard freshmen from relatively modest social backgrounds increased in stature by around 40 mm from 1930 to 1958, whereas those from wealthy backgrounds showed no change. Cameron (1979) has published data showing, very convincingly, that the secular trend for stature had levelled off, for children attending schools in the London area, by about 1960 (Figure 9.7). Tanner (1978) also showed that the secular decrease in

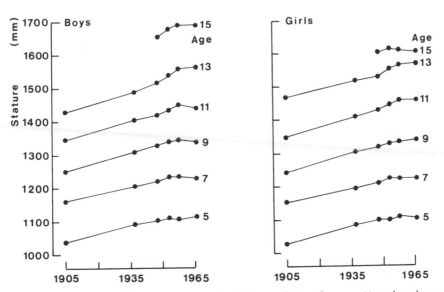

Figure 9.7 Secular trend in the average stature of children in the London area. Note that changes have been minimal since 1960. (Data from Cameron (1979).)

the age of menarche had come to a halt by about this time both in London and in Oslo. The subsequent national survey of Rona and Altman (1977) confirm the impression that in Great Britain the secular trend has now reached a steady state. Rona (1981) was prepared to conclude that 'there is no evidence that the secular trend in growth has continued after 1959 in the UK'. Similarly, Roche (1979) reported that national surveys of US children and youths in 1962 and 1974 show constancy of stature (except at the 5th and 10th %ile levels where small increases have occurred.) More recently, Chinn *et al.* (1989) have analysed the evidence for a continued secular trend in the heights of English and Scottish children over the years 1972–1986. The conclusion that they reached was that in the 5–11-year-old age group the secular trend has now ceased; and that the probability was that the upward trend in adult stature has likewise come to a halt.

Overall then, the consensus view of anthropometrists and human biologists, is that in the industrialized societies of Europe and North America, the upward secular trend in human stature has now come to a halt. Many people seem to find this surprising, however. The experiences of many schoolteachers, for example, tell them otherwise – and they are quite convinced in their own minds that the children they teach are still getting taller. It is difficult to account for this disparity between popular opinion and the available statistical evidence. In reality it is difficult to be absolutely certain either way. The differences that are known to exist between different parts of the UK and different social classes indicate that conclusions based upon small-scale or regional studies involving what may well be non-representative samples may very well be confusing.

If the secular trend has indeed come to a halt, then its absence admits of two possible interpretations. The optimistic is that conditions for growth have been optimized and that all children are now reaching their genetic ceilings. The pessimistic is that the percentage of children raised under optimal environmental conditions has ceased to increase. The continued existence of significant social class differences in growth (see below) tend towards the pessimistic interpretation.

Is the continuance or otherwise of the secular trend important in terms of practical ergonomics? If the upward trend in the stature of *children* is still increasing then it seems wholly unlikely that the rate of change can be great enough to invalidate the anthropometric assumptions upon which design standards for school furniture are currently based. But even if the upward trend in the stature of *young adults* has now ceased, the upward trend in the stature of the *adult populations as a whole* may well continue to increase into the first or second decade of the next century. It is difficult to predict the likely magnitude of these changes, since the effects of the secular trend are confounded with those of the ageing process *per se* and with demographic changes in the age structure of the population (see below). If the adult population of the next century had anthropometric characteristics similar to those of the 19–25-year-olds of the 1980s then the consequences of these changes could be significant in some areas of ergonomics practice – but only marginally so.

9.5   Social class and occupation

Social class and occupation are inextricably linked – so much so that the latter is generally used as an operational measure of the former. The widely used system of the Office of Population Censuses and Surveys, known as the 'Registrar-General's

Classification', divides occupations into six categories: I Professional; II Intermediate; IIIA Skilled Non-Manual; IIIB Skilled Manual; IV Semi-Skilled Manual and V Unskilled Manual. Every so often it is necessary to re-classify an occupation as its perceived status changes.

In a fascinating study of primiparae (women pregnant for the first time) in Aberdeen, Thomson (1959) found that stature was stratified by the occupation of the subject's father, by her own occupation, and by her husband's occupation but, most remarkably, that tall girls had a stronger tendency to marry upwards, with respect to their father's and their own occupations, than did short ones.

Social class differences in stature remain marked. Knight (1984), in a nationwide study of the adult population of Great Britain, found an average stature of 1755 mm for men and 1625 mm for women in social classes I and II as against 1723 and 1596 mm in social classes IV and V. The differences were of similar magnitude for all age cohorts. The pattern was less clear for body weight. The same survey also showed regional differences; ranging from an average stature of 1751 mm for men and 1619 mm for women in south-west England to 1719 and 1594 mm in Wales.

Extensive British data show class differences in the growth of schoolchildren. Rona (1981) has reviewed the evidence of British surveys over the last thirty years or so. A difference of between 10 and 20 mm in average stature between the top and bottom classes in the Registrar-General's table already exists at 2 years. By 7 years this has widened to 30–40 mm, a gap that has remained constant for the last thirty years. In the most recent of these surveys, Goldstein (1971) and Rona et al. (1978), the differences between classes I to IV were relatively modest, but it seems that at present differences in primary-school children are mainly due to those in social class V. Rona et al. (1978) showed that the children of unemployed fathers are shorter in stature, but that the parents of these children were also shorter within each social class. Lindgren (1976) reported an extensive survey of urban children in Sweden between 10 and 18 years of age. There was no difference in height, at any age, between the social classes as defined either by the father's occupation or family income. Sweden is the only country in the world where this is known to be the case – a fact that Tanner (1978) takes to be an operational and biological measure of the existence of a 'classless society'.

In some circumstances 'self-selection' may occur, with individuals gravitating towards occupations to which their physique is well suited. The physical content of the occupation itself may also exert a training effect – or, perhaps more importantly, a de-training effect in the case of sedentary lifestyles. (The most extreme examples of these effects are provided by athletes; see Wilmore 1976.)

A classic study of 'selection and de-training' is that of Morris et al. (1956), who investigated the waist and chest girths of the uniforms of London busmen – both drivers and conductors – ranging in age from 25 to 64 years. In addition to seeing a steady increase with age in both groups, the girths of the drivers were greater than those of the conductors – even in the youngest age group. The authors postulated therefore that 'the men have brought these differences into the jobs with them'. Passing from the ridiculous (if we may so term the busmen's trousers) to the almost indisputably sublime, two studies of ballerinas are worth mention. Grahame and Jenkins (1972) measured the joint flexibility of a group of female ballet students and found it to be greater than controls, even for joints, such as those of the little finger, which were not trained as such. The authors therefore concluded that only girls gifted with generalized joint flexibility would undertake the rigours of ballet train-

ing. Vincent (1979) has documented the appalling, and sometimes disastrous, lengths
to which these girls will sometimes go in 'competing with the sylph' – that is in
pursuit of the unnaturally slender body form which is sufficiently otherworldly for
their art.

9.6 Ageing

Figure 9.8 shows the average heights and weights of the adult civilian populations of
Great Britain and the US plotted against age. A steady decline in stature is apparent, whereas weight climbs steadily before subsequently declining at around 50 years
in men or 60 years in women. In analysing such a pattern we must consider the
combined effects of the ageing process and the secular trend, together with the possibilities of differential mortality (i.e. that people with certain kinds of physique may
tend to die younger.) Damon (1973) showed that men of average height and weight
had greater longevity than those who deviated strongly in either respect. These
interactions require multicohort longitudinal studies for their elucidation. Investigations of this kind include the Welsh study of Miall et al. (1967) and the extensive
Boston programme of the Veterans Administration (Damon et al. 1972, Friedlander
et al. 1977, Borkan et al. 1983.) Longitudinal studies show that at around 40 years of
age we begin to shrink in stature, that the shrinkage accelerates with age, and that
women shrink more than men. The shrinkage is generally believed to occur in the
intervertebral discs of the spine – resulting in the characteristic round back of the
elderly (e.g. Trotter and Gleser 1951) – although Borkan et al. (1983) suggest that
some decrease also occurs in the lower limbs. The data show a longitudinal increase
in weight for height until 55 years followed by a decline. Friedlander et al. (1977)
showed a steady longitudinal increase not only in hip breadth but also in the bi-iliac
breadth of the bony pelvis. The mechanism of the latter is obscure but it suggests
that 'middle-age spread' may not be totally due to the accumulation of fat, but may
also involve changes in the bony pelvis.

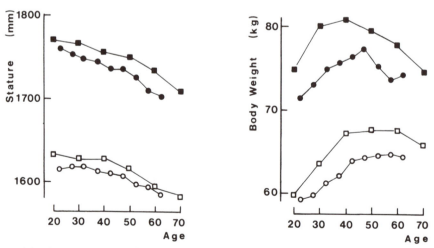

Figure 9.8   Average stature and weight in samples of adults of various ages. ■ = men, USA;
● = men, Britain; □ = women, USA; ○ = women, Britain.

If we assume that no secular change has occurred in the proportions of the body, then proportional ratios calculated from cross-sectional studies should be comparable with the longitudinal results. Figure 9.3 plotted from the data of Stoudt *et al.* (1965, 1970) shows that this is indeed the case. The proportional decrease in sitting height is compatible with the spinal shrinkage explanation of stature decline, and the greater change in women matches the longitudinal findings of Miall *et al.* (1967). Dimensions with a substantial soft-tissue component such as hip breadth and buttock–knee length show a proportional increase (until 75 years) which is more pronounced in women. The proportional decline in the biacromial breadth of men over 60 years old presumably reflects the characteristic rounding of the shoulders of the elderly. It is of interest that for both sitting height and biacromial breadth the ageing process finally abolishes the sex difference altogether.

Cross-sectional studies such as Stoudt *et al.* (1970) have shown an increase in skinfold thickness followed by a decline at around 40 years in men and 60 years in women. There is evidence, however, that this represents a redistribution rather than a loss of body fat. Durnin and Womersley (1974) showed that the relationship between skinfold thickness and whole body fat, as measured by densitometry, changes with age. It seems there is a transfer of fat from subcutaneous positions to deep ones (e.g. around the abdominal organs). The net quantity as a percentage of body weight continues to increase, and the longitudinal decline in weight that we see late in life is probably due, therefore, to the loss of lean tissue. Borkan and Norris (1977) found that the weight of fat was constant with age in a cross-sectional sample of middle-aged men, but that lean tissue declined markedly. Subcutaneous fat decreased on the trunk but increased on the hips, but this was accompanied by an increase of abdominal (waist) circumference indicative of a sagging of the abdominal contents (due presumably to increased internal fat and decreased muscular resistance.) A similar redistribution presumably occurs in women – but there is little numerical evidence.

The above studies are all based on the populations of the US and Great Britain, where obesity, consequent upon an abundant food supply and a sedentary lifestyle, is prevalent. The situation in other communities will, of course, be different, and in societies where food is scarce, adult increase in body weight does not occur.

The loss of lean body weight is due principally to a wasting away of muscles (although the bones also become less dense in later life) and this leads to a decrease in muscular strength – as shown for example in Figure 9.3.

According, for example, to Asmussen and Heeboll-Nielsen (1962), the decline is more rapid in women than in men; and more rapid in lower limb muscle groups than in upper limb muscle groups. In other words, women age more rapidly than men and the legs give out first. Both of these conclusions have been challenged, however: the first by Montoye and Lamphier (1977) and the second by Viitisalo *et al.* (1985).

We live in a 'greying' society. Figure 9.9 shows some recent demographic predictions. In 1971 about one person in six in the UK was of retirement age (i.e. 65 for men, 60 for women); by 2031, it is estimated, the figure will be closer to one person in four. The rate of increase is greatest in the oldest age groups – specifically the over-75s who will increase dramatically in numbers by the turn of the century.

Beyond the middle years of life, most of us will tend to suffer from a steady diminution in our functional capacities – due partly to the ageing process as such and partly to the effects of previous disease or injury from which recovery has been

BODYSPACE

incomplete. As a consequence, we experience a steady increase in the number of
critical mismatches that we encounter in the performance of the everyday tasks of
life. The net effect of these changes are illustrated in Figure 9.10 which shows the
percentage of people in different age groups having one or more specific disabilities:
that is, one or more functional impairments that lead to significant difficulties in the
performance of tasks in everyday life. The figure takes a dramatic upswing beyond
the age of 60. As things stand at present therefore, we typically seem to continue

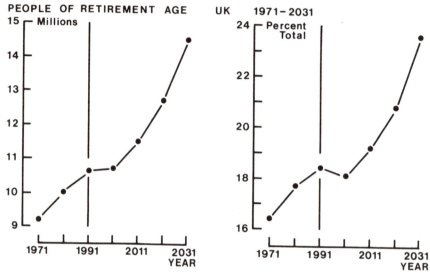

Figure 9.9   The ageing population: people of retirement age in the UK; (left) millions; (right)
percentage of the total population. Data from *Social Trends 20*, HMSO, 1990. (From S. Pheasant,
Ergonomics, Work and Health, Macmillan, 1991, Figure 16.1, p. 324, reproduced with kind
permission.)

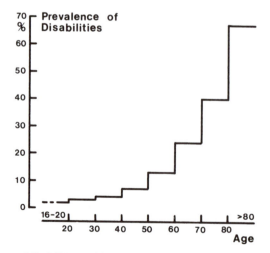

Figure 9.10   Prevalence of disabilities in different age groups. Data from Martin *et al.* (1988). (From
S. Pheasant, *Ergonomics, Work and Health*, Macmillan, 1991, Figure 16.2 p. 327, reproduced with
kind permission.)

working until we reach the age when our bodily framework starts to pack up on us. There is something of an irony in this.

The rate of onset of the decrepitude which comes with old age is highly variable. Part of this is just luck – a matter of the genes we are born with and the misfortunes we encounter along the way. Lifestyle is a major factor, however. In particular, it seems fairly certain that (within limits) regular physical activity can fend off the ageing process. Unfortunately, however, we not uncommonly get trapped in a downward spiral in which diminished functional capacity leads to a reduction of activity; which leads to a further reduction in functional capacity and so on. We encounter difficulties in doing things, so we stop doing them, and in due course are able to do less and less. The problems of the ageing society present a major challenge for ergonomics.

CHAPTER TEN

Anthropometric data

As we noted in an earlier chapter (see Section 2.4), there are two possible approaches to the business of assembling an anthropometric database: the purist and the pragmatic. In compiling the tables that follow I have adopted the latter. The technique used most extensively is that of ratio scaling, a more detailed discussion of which will be found in the Appendix and in Pheasant 1982a,b. Dimensions marked with an asterisk (*) are quoted from the original sources (M for men only, W for women only). The remainder have been estimated. Details of the sources are given below. Definitions of the dimensions which are tabulated together with some notes on application will also be found in Chapter 2.

The first table in this chapter (Table 10.1) is for the adult population of the UK. It is the same as Table 2.3 which, the reader will recall, is that of our 'standard reference population' used throughout the book, and is repeated here for ease of reference. This is followed by tables for three subsets of the British adult population, broken down by age, who are somewhat taller or shorter as the case may be. The first of these (Table 10.2 for 19–25-year-olds) may be regarded as a best estimate of the anthropometric characteristics of the British adult population in the early part of the next century. The reader will note that the actual figures in these tables do not in reality vary very greatly; and in practical terms the differences between them are of only marginal significance. They have been included so as to enable the reader to decide for herself the extent to which it may be necessary to take such differences into account in practical design work.

Tables 10.5 and 10.6 are two different estimates for the elderly population of the UK. The former is a more conservative estimate than the latter.

There follow four representative datasets for various European countries. Swedish men (Table 10.7) are much the same as British men; whereas Swedish women are a fair bit taller than British women and somewhat lighter in weight and more slender in their build. The Dutch (Table 10.8) are the tallest people in Europe (and the tallest included in the present database) in the case of the men by quite a fair margin. Table 10.9 is based upon a particularly thorough and well-conducted survey of French drivers reported by Rebiffe *et al.* 1983. On the assumption that these drivers were a representative sample of the adult population as a whole, then the French are a little shorter than the British. Taken together these various populations may be assumed to bracket the adult population of northern Europe. The populations of southern Europe will in general tend to be a little shorter than northern European populations. The data given for Polish industrial workers in Table 10.10 probably represent the shorter limit of European populations (or something very close to it).

US adults (Table 10.11) are a little taller and a little heavier than their UK counterparts, although again the difference will be of relatively little significance in practical terms.

The data given in Table 10.12 for male Brazilian industrial workers place them close to the bottom of the stature range of European populations. Asian populations (Tables 10.13–10.16) are shorter still, the shortest population in the present collection being the sample of male Indian industrial workers reported by Gite and Yadav (1989) which is given in Table 10.14.

The tables conclude with data for British children from birth to maturity.

10.1 Notes on sources of data

Each table is derived by a combination of two or more sources:

- a size source which gives us the mean and standard deviation for stature (or crown–heel length) in the relevant target population;
- one or more shape sources from which the coefficients E_1 and E_2 (as defined in equations A41 and A42) are calculated. The shape sources must be of the same age and sex as the target population and have a similar ethnic admixture.

10.2 British adults (Tables 10.1–10.6)

Stature data were taken from a survey of a nationwide stratified sample of households conducted in 1981 by the Office of Population Censuses and Surveys (OPCS 1981, Knight 1984). We may have considerable confidence in the validity and reliability of these data. Reference sources for E_1 and E_2 were: US civilians (Stoudt et al. 1965, 1970) for dimensions 8, 11, 13, 14, 15, 16, 17, 18, and 19; French drivers (Rebiffe et al. 1983) for dimensions 22, 23, 24, 25, and 36; British drivers (Haslegrave 1979) for dimensions 12, 20, and 21. The remaining dimensions were calculated from a variety of US military surveys published in NASA (1978). Separate E coefficients were established for the different age bands.

The over-65-year-olds presented a problem. The OPCS stature data only extend to 65 years. An alternative source would be the survey by the Institute of Consumer Ergonomics (1983) of the inmates of geriatric institutions. The latter were measured shod but if we subtract a nominal 20 mm for heels we still see that their stature is very much less than that of the 45–65-year-olds in the OPCS sample. In the light of this discrepancy two tables have been prepared. Table 10.5 is based on stature data estimated on the assumption that the decline in stature after 65 years is of a similar magnitude in Great Britain and the US (as documented by Stoudt et al. 1965). Table 10.6 is based on the Institute of Consumer Ergonomics (1983) survey – to which the reader should turn for further information.

10.3   Adult populations of other countries (Tables 10.7–10.16)

Table 10.7 is based on surveys of Swedish male and female workers by Lewin (1969) and Swedish women by Inglemark and Lewin (1968); Table 10.8 on a set of estimates kindly provided by Joban Molenbroek of Delft University of Technology; Table 10.9 on the survey of French drivers by Rebiffe et al. (1983); and Table 10.10 on a survey of Polish industrial workers reported by Batogowska and Slowikowski (1974). Table 10.11 is based upon a major survey of US adults conducted in 1971–1974 and deemed to be the most up to date available (Abraham 1979). E coefficients were in all cases the same as for Table 10.1.

Table 10.12 is based upon a publication of the Instituto Nacional de Tecnologia (1989) in Rio; Table 10.3 on Abeysekera and Shanavaz (1987); and Table 10.14 on Gite and Yadav (1989). In each case missing data were, wherever possible, estimated by scaling up from the nearest available dimension; and failing that, the E coefficients of Table 10.1 were used.

Stature data and E coefficients for men used in Table 10.16 are based on sources cited in NASA (1978); E coefficients for women based on the assumption that differences in proportion between Japanese men and women are similar to those for European men and women. (In the absence of suitable reference data, estimates were made by scaling from the nearest available dimension.)

Table 10.15 is based on data from a hitherto unpublished survey of Chinese industrial workers in Hong Kong, kindly supplied by Bill Evans of the Department of Industrial Engineering at the University of Hong Kong. Dimensions marked with an asterisk (*) are quoted directly; the remainder are estimated as for Table 10.16.

10.4   Infants (Tables 10.17–10.21)

The crown–heel length data were taken from Tanner et al. (1966). Male and female data were combined using equations A23 and A24 and the 'point in time' values of the original were converted to 'period of time' values using the equation of Healy (1962). Values of E_1 and E_2 were taken from the survey of US infants published by Snyder et al. (1977) – linear interpolation was required to adjust the data to appropriate mid-sample ages.

10.5   Children and youths (Tables 10.22–10.38)

In these tables a 5-year-old, for example, is anyone between their fifth and sixth birthdays. The stature data employed were taken from a major survey of British schoolchildren published by the Department of Education and Science (1972). Shape data were from Martin (1960) and Snyder et al. (1977). The predictions were edited to ensure steady unidirectional growth in all percentiles and compatibility with young adult data. The data are the same as those of the Department of Education and Science (1985) except that a different selection of anthropometric variables are included – additional details of the editing may be found there. The 2-year-old data presented a particular problem since no suitable sources existed. It was therefore assumed that 2- and 3-year-olds differed by the same amount as 3- and 4-year-olds. No suitable sources exist for the chest or abdominal depths of the under-18-year-olds. E coefficients were calculated from Snyder *et al.* (1977) for equivalent circumferences; they were then scaled down according to young adult depth data. Relative chest depth for girls was assumed to be the same as that for boys until 11 years and then to approach the adult female proportions by steady annual increments.

Subsequent to the preparation of these estimates a British Standard has been published dealing with the body measurements of children (BS 7231). Most of the dimensions given in this standard are such as to be appropriate for clothing design rather than other areas of application. The stature data of this standard differ somewhat from those of the DES survey on which the present estimates are based. I have no means of knowing which is the more representative survey. The reader is invited to scale up the present estimates using the stature data given in the standard, should she consider it appropriate to do so.

The anthropometric tables

See over for anthropometric database.

Table 10.1 Anthropometric estimates for British adults aged 19–65 years (all dimensions in mm, except for body weight, given in kg).

Dimension	Men				Women			
	5th %ile	50th %ile	95th %ile	SD	5th %ile	50th %ile	95th %ile	SD
1. Stature	1625	1740	1855	70	1505	1610	1710	62*
2. Eye height	1515	1630	1745	69	1405	1505	1610	61
3. Shoulder height	1315	1425	1535	66	1215	1310	1405	58
4. Elbow height	1005	1090	1180	52	930	1005	1085	46
5. Hip height	840	920	1000	50	740	810	885	43
6. Knuckle height	690	755	825	41	660	720	780	36
7. Fingertip height	590	655	720	38	560	625	685	38
8. Sitting height	850	910	965	36	795	850	910	35
9. Sitting eye height	735	790	845	35	685	740	795	33
10. Sitting shoulder height	540	595	645	32	505	555	610	31
11. Sitting elbow height	195	245	295	31	185	235	280	29
12. Thigh thickness	135	160	185	15	125	155	180	17
13. Buttock–knee length	540	595	645	31	520	570	620	30
14. Buttock–popliteal length	440	495	550	32	435	480	530	30
15. Knee height	490	545	595	32	455	500	540	27
16. Popliteal height	395	440	490	29	355	400	445	27
17. Shoulder breadth (bideltoid)	420	465	510	28	355	395	435	24
18. Shoulder breadth (biacromial)	365	400	430	20	325	355	385	18
19. Hip breadth	310	360	405	29	310	370	435	38
20. Chest (bust) depth	215	250	285	22	210	250	295	27
21. Abdominal depth	220	270	325	32	205	255	305	30
22. Shoulder–elbow length	330	365	395	20	300	330	360	17
23. Elbow–fingertip length	440	475	510	21	400	430	460	19
24. Upper limb length	720	780	840	36	655	705	760	32
25. Shoulder–grip length	610	665	715	32	555	600	650	29
26. Head length	180	195	205	8	165	180	190	7
27. Head breadth	145	155	165	6	135	145	150	6
28. Hand length	175	190	205	10	160	175	190	9
29. Hand breadth	80	85	95	5	70	75	85	4
30. Foot length	240	265	285	14	215	235	255	12
31. Foot breadth	85	95	110	6	80	90	100	6
32. Span	1655	1790	1925	83	1490	1605	1725	71
33. Elbow span	865	945	1020	47	780	850	920	43
34. Vertical grip reach (standing)	1925	2060	2190	80	1790	1905	2020	71
35. Vertical grip reach (sitting)	1145	1245	1340	60	1060	1150	1235	53
36. Forward grip reach	720	780	835	34	650	705	755	31
Body weight	*55*	*75*	*94*	*12*	*44*	*63*	*81*	*11**

See notes on pp. 175–6.

Table 10.2 Anthropometric estimates for British adults aged 19–25 years (all dimensions in mm).

Dimension	Men 5th %ile	50th %ile	95th %ile	SD	Women 5th %ile	50th %ile	95th %ile	SD
1. Stature	1640	1760	1880	73	1520	1620	1720	61*
2. Eye height	1530	1650	1770	72	1415	1515	1615	60
3. Shoulder height	1330	1445	1555	69	1225	1320	1410	57
4. Elbow height	1020	1105	1195	54	940	1015	1090	45
5. Hip height	850	935	1020	52	745	815	885	43
6. Knuckle height	695	765	835	42	665	725	785	35
7. Fingertip height	595	665	730	40	565	630	690	38
8. Sitting height	855	915	980	37	800	855	915	35
9. Sitting eye height	740	795	855	36	690	745	800	33
10. Sitting shoulder height	545	600	655	33	510	560	610	31
11. Sitting elbow height	195	245	300	32	180	230	275	28
12. Thigh thickness	130	160	185	16	120	150	175	16
13. Buttock–knee length	545	595	650	32	520	565	615	29
14. Buttock–popliteal length	445	500	555	34	430	475	525	29
15. Knee height	495	550	605	33	460	500	545	26
16. Popliteal height	400	445	495	30	355	400	445	27
17. Shoulder breadth (bideltoid)	415	465	510	29	355	395	435	24
18. Shoulder breadth (biacromial)	370	405	440	21	330	360	390	18
19. Hip breadth	300	350	400	31	300	350	400	29
20. Chest (bust) depth	185	225	270	26	190	235	275	26
21. Abdominal depth	195	240	280	26	185	220	260	22
22. Shoulder–elbow length	335	370	405	21	305	330	360	17
23. Elbow–fingertip length	445	480	515	22	400	430	465	19
24. Upper limb length	730	790	850	37	660	710	760	32
25. Shoulder–grip length	615	670	730	34	560	605	650	29
26. Head length	185	195	210	8	170	180	190	7
27. Head breadth	145	155	165	7	135	145	155	5
28. Hand length	175	190	210	10	160	175	190	9
29. Hand breadth	80	90	95	5	70	75	85	4
30. Foot length	245	270	290	15	220	240	260	12
31. Foot breadth	90	100	110	7	80	90	100	5
32. Span	1670	1815	1955	86	1500	1615	1730	70
33. Elbow span	875	955	1035	49	785	855	925	42
34. Vertical grip reach (standing)	1950	2085	2220	83	1805	1915	2030	70
35. Vertical grip reach (sitting)	1155	1260	1360	63	1070	1155	1245	52
36. Forward grip reach	730	790	845	36	655	705	755	31

Note: Best estimate for overall British population in the year 2000.
See also notes on pp. 175–6.

Table 10.3 Anthropometric estimates for British adults aged 19–45 years (all dimensions in mm).

	Men				Women			
Dimension	5th %ile	50th %ile	95th %ile	SD	5th %ile	50th %ile	95th %ile	SD
1. Stature	1635	1745	1860	69	1515	1615	1715	61*
2. Eye height	1525	1635	1750	68	1415	1515	1615	60
3. Shoulder height	1325	1435	1540	65	1225	1315	1410	57
4. Elbow height	1015	1100	1185	51	940	1015	1085	45
5. Hip height	845	925	1005	49	745	815	885	43
6. Knuckle height	695	760	825	40	665	725	780	35
7. Fingertip height	595	660	720	38	565	625	690	38
8. Sitting height	855	915	970	35	800	855	915	35
9. Sitting eye height	740	795	850	34	690	745	800	33
10. Sitting shoulder height	545	595	650	31	510	560	610	31
11. Sitting elbow height	195	245	295	30	190	235	280	28
12. Thigh thickness	135	160	185	15	125	155	180	16
13. Buttock–knee length	545	595	645	30	520	570	610	29
14. Buttock–popliteal length	445	495	550	32	435	480	530	29
15. Knee height	495	545	600	31	460	500	545	26
16. Popliteal height	395	445	490	28	355	400	445	27
17. Shoulder breadth (bideltoid)	420	465	510	28	355	395	435	24
18. Shoulder breadth (biacromial)	365	400	435	20	330	360	390	18
19. Hip breadth	310	355	405	29	300	365	425	37
20. Chest (bust) depth	200	240	275	23	195	240	285	26
21. Abdominal depth	210	255	300	28	195	245	290	29
22. Shoulder–elbow length	330	365	400	20	305	330	360	17
23. Elbow–fingertip length	445	475	510	21	400	430	460	19
24. Upper limb length	725	785	840	35	655	710	760	32
25. Shoulder–grip length	615	665	720	32	555	605	650	29
26. Head length	185	195	210	8	165	180	190	7
27. Head breadth	145	155	165	6	135	145	155	5
28. Hand length	175	190	205	10	160	175	190	9
29. Hand breadth	80	85	95	5	70	75	85	4
30. Foot length	245	265	290	14	215	235	255	12
31. Foot breadth	90	100	110	6	80	90	100	5
32. Span	1665	1800	1935	81	1500	1615	1730	70
33. Elbow span	875	950	1025	46	785	855	920	42
34. Vertical grip reach (standing)	1940	2070	2200	79	1800	1915	2030	70
35. Vertical grip reach (sitting)	1150	1250	1345	59	1070	1155	1240	52
36. Forward grip reach	725	785	840	34	655	705	755	31

See notes on pp. 175–6.

Table 10.4 Anthropometric estimates for British adults aged 45–65 years (all dimensions in mm).

	Men				Women			
Dimension	5th %ile	50th %ile	95th %ile	SD	5th %ile	50th %ile	95th %ile	SD
1. Stature	1610	1720	1830	67	1495	1595	1695	61*
2. Eye height	1505	1610	1720	66	1395	1495	1595	60
3. Shoulder height	1305	1410	1515	63	1205	1300	1395	57
4. Elbow height	1000	1080	1160	50	925	1000	1075	45
5. Hip height	835	910	990	48	735	805	875	43
6. Knuckle height	685	750	810	39	655	715	775	35
7. Fingertip height	590	650	710	37	555	620	680	38
8. Sitting height	840	900	955	34	785	845	900	35
9. Sitting eye height	725	780	835	34	680	735	790	33
10. Sitting shoulder height	535	585	635	30	500	550	600	31
11. Sitting elbow height	190	240	290	29	185	230	280	28
12. Thigh thickness	135	160	185	15	130	155	180	16
13. Buttock–knee length	540	585	635	29	520	570	620	29
14. Buttock–popliteal length	440	490	540	31	435	480	530	29
15. Knee height	490	540	590	30	450	495	540	26
16. Popliteal height	390	435	480	27	350	395	440	27
17. Shoulder breadth (bideltoid)	415	460	505	27	350	390	430	24
18. Shoulder breadth (biacromial)	360	395	425	19	325	355	385	18
19. Hip breadth	310	360	405	28	315	375	440	37
20. Chest (bust) depth	225	260	295	20	220	265	305	26
21. Abdominal depth	230	285	340	34	220	270	320	31
22. Shoulder–elbow length	330	360	390	19	300	325	355	17
23. Elbow–fingertip length	435	470	505	20	395	425	455	19
24. Upper limb length	715	770	830	34	650	700	755	32
25. Shoulder–grip length	605	655	710	31	550	595	645	29
26. Head length	180	195	205	7	165	175	190	7
27. Head breadth	140	150	160	6	135	140	150	5
28. Hand length	170	185	205	9	155	170	185	9
29. Hand breadth	80	85	95	5	70	75	80	4
30. Foot length	240	260	285	13	215	235	255	12
31. Foot breadth	85	95	105	6	80	90	95	5
32. Span	1640	1770	1900	79	1480	1595	1710	70
33. Elbow span	860	935	1010	45	775	845	910	42
34. Vertical grip reach (standing)	1910	2035	2160	76	1775	1890	2000	70
35. Vertical grip reach (sitting)	1135	1230	1325	58	1055	1140	1225	52
36. Forward grip reach	715	770	825	33	645	695	750	31

See notes on pp. 175–6.

Table 10.5 Anthropometric estimates for British adults aged 65–80 years (all dimensions in mm).

Dimension	Men				Women			
	5th %ile	50th %ile	95th %ile	SD	5th %ile	50th %ile	95th %ile	SD
1. Stature	1575	1685	1790	66	1475	1570	1670	60
2. Eye height	1470	1575	1685	65	1375	1475	1570	59
3. Shoulder height	1280	1380	1480	62	1190	1280	1375	56
4. Elbow height	975	1055	1135	49	910	985	1055	44
5. Hip height	820	895	975	47	740	810	875	42
6. Knuckle height	670	730	795	38	645	705	760	35
7. Fingertip height	575	635	695	36	550	610	670	37
8. Sitting height	815	875	930	36	750	815	885	41
9. Sitting eye height	705	760	815	34	645	710	770	38
10. Sitting shoulder height	520	570	625	32	475	535	590	36
11. Sitting elbow height	175	220	270	29	165	210	260	28
12. Thigh thickness	125	150	175	15	115	145	170	16
13. Buttock–knee length	530	580	625	29	520	565	615	29
14. Buttock–popliteal length	430	485	535	31	430	480	525	29
15. Knee height	480	525	575	30	455	500	540	26
16. Popliteal height	385	425	470	27	355	395	440	26
17. Shoulder breadth (bideltoid)	400	445	485	26	345	385	425	23
18. Shoulder breadth (biacromial)	350	375	405	17	320	350	380	17
19. Hip breadth	305	350	395	28	310	370	430	37
20. Chest (bust) depth	225	260	290	20	220	265	305	26
21. Abdominal depth	245	300	355	33	225	270	320	30
22. Shoulder–elbow length	320	350	385	19	295	320	350	17
23. Elbow–fingertip length	425	460	490	20	390	420	450	19
24. Upper limb length	700	755	810	34	640	690	740	31
25. Shoulder–grip length	595	645	695	30	540	590	635	28
26. Head length	175	190	200	7	165	175	185	7
27. Head breadth	140	150	160	6	130	140	150	5
28. Hand length	170	185	200	9	155	170	185	9
29. Hand breadth	75	85	90	5	65	75	80	4
30. Foot length	235	255	280	13	210	230	250	12
31. Foot breadth	85	95	105	6	80	85	95	5
32. Span	1605	1735	1860	78	1460	1570	1685	68
33. Elbow span	840	915	985	44	760	830	900	41
34. Vertical grip reach (standing)	1840	1965	2090	75	1725	1835	1950	68
35. Vertical grip reach (sitting)	1110	1205	1295	57	1040	1125	1210	52
36. Forward grip reach	700	755	805	32	640	685	735	30

See notes on pp. 175–6.

Table 10.6 Anthropometric estimates for 'elderly people' (all dimensions in mm).

Dimension	Men				Women			
	5th %ile	50th %ile	95th %ile	SD	5th %ile	50th %ile	95th %ile	SD
1. Stature	1515	1640	1765	77	1400	1515	1630	70*
2. Eye height	1410	1535	1660	76	1305	1420	1535	69
3. Shoulder height	1225	1345	1465	72	1130	1235	1340	65
4. Elbow height	935	1025	1120	57	860	945	1030	52
5. Hip height	785	875	965	55	700	780	860	49
6. Knuckle height	640	715	785	45	610	680	745	41
7. Fingertip height	550	620	690	42	515	590	660	43
8. Sitting height	785	850	920	42	710	785	865	48
9. Sitting eye height	675	740	805	40	610	685	755	45
10. Sitting shoulder height	495	555	615	37	445	515	585	42
11. Sitting elbow height	160	215	270	34	150	205	255	32
12. Thigh thickness	120	145	175	17	105	140	170	19
13. Buttock–knee length	510	565	620	34	490	545	600	34
14. Buttock–popliteal length	410	470	530	36	405	460	515	34
15. Knee height	455	515	570	35	430	480	530	30
16. Popliteal height	365	415	470	32	330	380	430	31
17. Shoulder breadth (bideltoid)	380	430	480	31	325	370	415	27
18. Shoulder breadth (biacromial)	335	365	400	20	305	335	370	20
19. Hip breadth	290	340	395	32	285	355	425	43
20. Chest (bust) depth	215	255	290	23	205	255	305	30
21. Abdominal depth	230	290	355	39	205	260	320	35
22. Shoulder–elbow length	305	345	380	22	280	310	345	20
23. Elbow–fingertip length	410	450	485	23	370	405	440	22
24. Upper limb length	670	735	800	39	605	665	725	36
25. Shoulder–grip length	570	625	685	35	510	565	620	33
26. Head length	170	185	200	8	155	170	180	8
27. Head breadth	135	145	155	7	125	135	145	6
28. Hand length	160	180	195	11	145	165	180	10
29. Hand breadth	75	80	90	5	65	70	80	5
30. Foot length	225	250	275	15	200	225	245	14
31. Foot breadth	80	90	105	7	75	85	95	6
32. Span	1540	1690	1840	91	1380	1515	1645	80
33. Elbow span	805	890	975	52	720	800	880	48
34. Vertical grip reach (standing)	1770	1915	2060	88	1640	1770	1900	80
35. Vertical grip reach (sitting)	1065	1175	1280	66	985	1085	1180	60
36. Forward grip reach	675	735	795	38	605	660	720	35

See notes on pp. 175–6.

Table 10.7 Anthropometric estimates for Swedish adults (all dimensions in mm, except for body weight, given in kg).

Dimension	Men				Women			
	5th %ile	50th %ile	95th %ile	SD	5th %ile	50th %ile	95th %ile	SD
1. Stature	1630	1740	1850	68	1540	1640	1740	62*
2. Eye height	1520	1630	1740	68	1435	1535	1635	62*
3. Shoulder height	1345	1445	1545	62	1255	1355	1455	60*
4. Elbow height	1020	1100	1180	49	905	1025	1145	73*
5. Hip height	815	890	965	45	745	830	915	52*
6. Knuckle height	720	760	800	25	675	735	795	36
7. Fingertip height	595	655	715	37	570	635	700	38
8. Sitting height	830	900	970	43	805	860	915	33*
9. Sitting eye height	715	785	855	42	705	755	805	30*
10. Sitting shoulder height	545	600	655	34	525	575	625	30*
11. Sitting elbow height	175	225	275	31	165	215	265	31*
12. Thigh thickness	120	152	180	18	130	155	180	16*
13. Buttock–knee length	545	595	645	30	525	585	645	35*
14. Buttock–popliteal length	430	480	530	30	430	485	540	33*
15. Knee height	480	530	580	30	455	500	545	28*
16. Popliteal height	385	430	475	27	350	400	450	29
17. Shoulder breadth (bideltoid)	420	465	510	27	355	390	425	20
18. Shoulder breadth (biacromial)	365	400	435	20	325	350	375	15*(W)
19. Hip breadth	310	360	410	29	315	365	415	31
20. Chest (bust) depth	185	220	255	21	185	241	300	35
21. Abdominal depth	190	240	290	31	180	245	310	40*
22. Shoulder–elbow length	330	365	400	20	305	335	365	17
23. Elbow–fingertip length	440	475	510	20	160	175	190	10*(W)
24. Upper limb length	720	780	840	35	660	705	750	28*(W)
25. Shoulder–grip length	615	665	715	31	555	595	635	24
26. Head length	185	195	205	7	170	180	190	7
27. Head breadth	145	155	165	6	135	145	155	6
28. Hand length	175	190	205	10	165	180	195	10*(W)
29. Hand breadth	75	85	95	5	70	75	80	4*(W)
30. Foot length	240	265	290	14	225	245	265	11*(W)
31. Foot breadth	85	95	105	6	85	95	105	7*(W)
32. Span	1660	1790	1920	80	1525	1640	1755	71
33. Elbow span	870	945	1020	45	795	865	935	43
34. Vertical grip reach (standing)	1930	2060	2190	78	1825	1940	2055	70
35. Vertical grip reach (sitting)	1150	1245	1340	58	1090	1175	1260	53
36. Forward grip reach	725	780	835	33	665	715	765	31
Body weight					48	59	70	7*

See notes on pp. 176.

Table 10.8 Anthropometric estimates for Dutch adults aged 20–60 (all dimensions in mm, except for body weight, given in kg).

	Men				Women			
Dimension	5th %ile	50th %ile	95th %ile	SD	5th %ile	50th %ile	95th %ile	SD
1. Stature	1690	1795	1900	65	1545	1650	1755	65*
2. Eye height	1575	1670	1765	59	1435	1530	1625	59*
3. Shoulder height	1400	1495	1590	58	1265	1365	1465	61*
4. Elbow height	1055	1135	1215	48	980	1050	1120	43*
5. Hip height	885	960	1035	46	780	845	910	40*
6. Knuckle height	745	795	845	30	705	775	845	42*
7. Fingertip height	645	690	735	28	605	675	745	42
8. Sitting height	885	940	995	34	820	875	930	33*
9. Sitting eye height	770	820	875	32	695	750	805	32*
10. Sitting shoulder height	570	620	670	31	515	565	615	30
11. Sitting elbow height	195	240	280	26	200	240	280	26*
12. Thigh thickness	120	140	160	12	125	150	175	17*
13. Buttock–knee length	575	620	665	28	550	600	650	31*
14. Buttock–popliteal length	470	520	570	30	440	495	550	32*
15. Knee height	520	565	610	28	450	505	560	32
16. Popliteal height	415	455	495	25	370	405	445	25*
17. Shoulder breadth (bideltoid)	430	475	520	28	355	400	445	27
18. Shoulder breadth (biacromial)	385	410	445	20	330	360	390	20*
19. Hip breadth	340	375	410	20	340	395	450	34
20. Chest (bust) depth	240	285	330	26	230	290	350	36*
21. Abdominal depth	245	310	375	38	230	295	360	40
22. Shoulder–elbow length	340	375	405	19	305	335	365	18
23. Elbow–fingertip length	455	490	525	20	405	440	475	20
24. Upper limb length	750	805	860	33	665	720	775	34
25. Shoulder–grip length	635	685	735	30	560	610	660	30
26. Head length	190	200	210	7	175	185	195	7
27. Head breadth	150	160	170	6	140	150	160	6
28. Hand length	180	195	210	9	160	175	190	9*
29. Hand breadth	80	90	100	5	70	80	90	4
30. Foot length	255	275	295	13	220	240	260	13
31. Foot breadth	90	100	110	6	80	90	100	6
32. Span	1720	1845	1970	77	1510	1635	1760	75
33. Elbow span	905	975	1045	44	790	865	920	45
34. Vertical grip reach (standing)	2000	2125	2250	76	1780	1905	2030	76*
35. Vertical grip reach (sitting)	1190	1280	1890	55	1085	1175	1265	56
36. Forward grip reach	680	745	810	38	635	705	780	44*
Body weight	*60*	*76*	*92*	*10*	*49*	*65*	*81*	*10**

See notes on p. 176.

Table 10.9 Anthropometric estimates for French drivers (all dimensions in mm, except for body weight, given in kg).

Dimension	Men				Women			
	5th %ile	50th %ile	95th %ile	SD	5th %ile	50th %ile	95th %ile	SD
1. Stature	1600	1715	1830	69	1500	1600	1700	61*
2. Eye height	1450	1560	1670	68	1400	1500	1600	60
3. Shoulder height	1300	1405	1510	65	1210	1305	1400	57
4. Elbow height	995	1080	1165	51	925	1000	1075	45
5. Hip height	815	895	975	49	750	820	890	43
6. Knuckle height	680	745	810	40	655	715	775	35
7. Fingertip height	580	645	710	38	560	620	680	37
8. Sitting height	850	910	970	35	810	860	910	31*
9. Sitting eye height	735	795	855	35	700	750	800	30*
10. Sitting shoulder height	570	620	670	31	535	580	625	27*
11. Sitting elbow height	190	240	290	30	185	230	275	28
12. Thigh thickness	150	180	210	17	135	165	195	17*
13. Buttock–knee length	550	595	640	28	520	565	610	28*
14. Buttock–popliteal length	435	480	525	26	415	460	505	26*
15. Knee height	485	530	575	26	455	495	535	24*
16. Popliteal height	385	425	465	25	350	390	430	23*
17. Shoulder breadth (bideltoid)	425	470	515	26	380	425	470	27*
18. Shoulder breadth (biacromial)	360	395	430	20	325	355	385	18
19. Hip breadth	330	370	410	24	330	380	430	30*
20. Chest (bust) depth	210	245	280	22	205	250	295	26
21. Abdominal depth	220	270	320	31	205	255	305	30
22. Shoulder–elbow length	325	360	395	20	300	330	360	17*
23. Elbow–fingertip length	435	470	505	21	395	425	455	19*
24. Upper limb length	710	770	830	35	650	705	760	32*
25. Shoulder–grip length	600	655	710	32	550	600	650	29
26. Head length	175	190	205	8	170	180	190	7
27. Head breadth	140	150	160	6	130	140	150	5
28. Hand length	170	185	200	10	160	175	190	9
29. Hand breadth	75	85	95	5	70	75	80	4
30. Foot length	235	260	285	14	215	235	255	12
31. Foot breadth	85	95	105	6	80	90	100	5
32. Span	1630	1765	1900	81	1485	1600	1715	70
33. Elbow span	855	930	1005	46	775	845	915	42
34. Vertical grip reach (standing)	1900	2030	2160	79	1780	1895	2010	70
35. Vertical grip reach (sitting)	1130	1225	1320	59	1060	1145	1230	52
36. Forward grip reach	715	770	825	34	645	700	755	32
Body weight	*58*	*73*	*95*	*11*	*47*	*58*	*78*	*10**

See notes on p. 176.

Table 10.10 Anthropometric estimates for Polish industrial workers (all dimensions in mm).

Dimension	Men				Women			
	5th %ile	50th %ile	95th %ile	SD	5th %ile	50th %ile	95th %ile	SD
1. Stature	1595	1695	1795	61	1480	1575	1670	58*
2. Eye height	1505	1600	1695	58	1390	1485	1580	57*
3. Shoulder height	1275	1365	1455	54	1170	1280	1390	68*
4. Elbow height	990	1065	1140	45	915	985	1055	43
5. Hip height	810	880	950	43	745	810	875	41
6. Knuckle height	545	595	640	29	535	570	605	22*
7. Fingertip height	615	675	735	36	590	645	700	32*
8. Sitting height	830	885	940	34	770	825	880	33*
9. Sitting eye height	720	780	840	36	665	725	785	35*
10. Sitting shoulder height	555	605	655	31	515	565	615	31*
11. Sitting elbow height	195	240	285	27	185	230	275	27
12. Thigh thickness	110	140	170	19	115	140	165	16*
13. Buttock–knee length	540	585	630	26	515	565	615	29*
14. Buttock–popliteal length	405	455	505	29	360	450	540	54*
15. Knee height	485	530	575	27	445	485	525	24*
16. Popliteal height	410	445	480	21	390	420	450	19*
17. Shoulder breadth (bideltoid)	405	440	475	21	350	380	410	18*
18. Shoulder breadth (biacromial)	360	390	420	18	320	350	380	17*
19. Hip breadth	305	345	385	25	315	360	405	26*
20. Chest (bust) depth	215	245	275	19	205	245	285	25
21. Abdominal depth	220	265	310	27	205	250	295	28
22. Shoulder–elbow length	310	330	350	13	280	300	320	12*
23. Elbow–fingertip length	430	460	490	18	390	420	450	18
24. Upper limb length	705	755	805	30	655	700	745	28*
25. Shoulder–grip length	640	675	710	22	595	625	655	18*
26. Head length	175	185	195	7	170	180	190	5*
27. Head breadth	145	155	165	6	140	150	160	5*
28. Hand length	175	190	205	8	160	175	190	8*
29. Hand breadth	80	90	100	5	75	80	85	4*
30. Foot length	240	260	280	12	210	230	250	11
31. Foot breadth	85	95	105	5	75	85	95	5
32. Span	1640	1755	1870	70	1505	1610	1715	65*
33. Elbow span	795	860	925	38	720	785	850	41*
34. Vertical grip reach (standing)	2065	2205	2345	84	1875	2005	2135	79*
35. Vertical grip reach (sitting)	1210	1290	1370	50	1115	1185	1255	42*
36. Forward grip reach	730	795	860	38	680	735	790	34*

See notes on p. 176.

Table 10.11 Anthropometric estimates for US adults aged 19–65 years (all dimensions in mm, except body weight, given in kg).

Dimension	Men 5th %ile	50th %ile	95th %ile	SD	Women 5th %ile	50th %ile	95th %ile	SD
1. Stature	1640	1755	1870	71	1520	1625	1730	64*
2. Eye height	1595	1710	1825	70	1420	1525	1630	63
3. Shoulder height	1330	1440	1550	67	1225	1325	1425	60
4. Elbow height	1020	1105	1190	53	945	1020	1095	47
5. Hip height	835	915	995	50	760	835	910	45
6. Knuckle height	700	765	830	41	670	730	790	37
7. Fingertip height	595	660	725	39	565	630	695	40
8. Sitting height	855	915	975	36	800	860	920	36
9. Sitting eye height	740	800	860	35	690	750	810	35
10. Sitting shoulder height	545	600	655	32	510	565	620	32
11. Sitting elbow height	195	245	295	31	185	235	285	29
12. Thigh thickness	135	160	185	16	125	155	185	17
13. Buttock–knee length	550	600	650	31	525	575	625	31
14. Buttock–popliteal length	445	500	555	33	440	490	540	31
15. Knee height	495	550	605	32	460	505	550	28
16. Popliteal height	395	445	495	29	360	405	450	28
17. Shoulder breadth (bideltoid)	425	470	515	28	360	400	440	25
18. Shoulder breadth (biacromial)	365	400	435	21	330	360	390	19
19. Hip breadth	310	360	410	30	310	375	440	39
20. Chest (bust) depth	220	255	290	22	210	255	300	28
21. Abdominal depth	220	275	330	32	210	260	310	31
22. Shoulder–elbow length	330	365	400	21	305	335	365	18
23. Elbow–fingertip length	445	480	515	21	400	435	470	20
24. Upper limb length	730	790	850	36	655	715	775	35
25. Shoulder–grip length	615	670	725	33	560	610	660	30
26. Head length	180	195	210	8	165	180	195	8
27. Head breadth	145	155	165	6	135	145	155	6
28. Hand length	175	191	205	10	160	175	190	10
29. Hand breadth	80	90	100	5	65	75	85	5
30. Foot length	240	265	290	14	220	240	260	13
31. Foot breadth	90	100	110	6	80	90	100	6
32. Span	1670	1810	1950	84	1505	1625	1745	73
33. Elbow span	875	955	1035	48	790	860	930	44
34. Vertical grip reach (standing)	1950	2080	2210	80	1805	1925	2045	73
35. Vertical grip reach (sitting)	1155	1255	1355	61	1070	1160	1250	55
36. Forward grip reach	725	785	845	35	655	710	765	32
Body weight	*55*	*78*	*102*	*14*	*41*	*65*	*89*	*15**

See notes on p. 176.

Table 10.12 Anthropometric estimates for Brazilian industrial workers (all dimensions in mm, except for body weight, given in kg).

	Men			
	5th %ile	50th %ile	95th %ile	SD
1. Stature	1595	1700	1810	66*
2. Eye height	1490	1595	1700	66*
3. Shoulder height	1315	1410	1510	60*
4. Elbow height	965	1045	1120	49*
5. Hip height	800	880	960	47
6. Knuckle height	655	720	785	40
7. Fingertip height	565	625	690	37*
8. Sitting height	825	880	940	35*
9. Sitting eye height	720	775	830	34*
10. Sitting shoulder height	550	595	645	29*
11. Sitting elbow height	185	230	275	28*
12. Thigh thickness	120	150	180	16*
13. Buttock–knee length	550	595	650	30*
14. Buttock–popliteal length	435	480	530	29*
15. Knee height	490	530	575	27*
16. Popliteal height	390	425	465	24*
17. Shoulder breadth (bideltoid)	400	445	490	27*
18. Shoulder breadth (biacromial)	355	385	415	19
19. Hip breadth	305	340	385	25*
20. Chest depth	205	235	275	22*
21. Abdominal depth	220	245	305	33*
22. Shoulder–elbow length	335	365	405	21*
23. Elbow–fingertip length	440	475	510	22
24. Upper limb length	725	785	850	38*
25. Shoulder–grip length	615	670	725	34
26. Head length	175	190	205	8
27. Head breadth	140	150	160	6
28. Hand length	170	185	200	9
29. Hand breadth	75	85	95	5
30. Foot length	240	260	280	12*
31. Foot breadth	95	100	110	5*
32. Span	1625	1755	1885	78
33. Elbow span	855	925	995	44
34. Vertical grip reach (standing)	1895	2020	2145	75
35. Vertical grip reach (sitting)	1130	1220	1310	56
36. Forward grip reach	710	765	820	32
Body weight	*52*	*66*	*86*	*11**

See notes on p. 176.

Table 10.13 Anthropometric estimates for Sri Lankan workers (all dimensions in mm, except for body weight, given in kg).

	Men				Women			
	5th %ile	50th %ile	95th %ile	SD	5th %ile	50th %ile	95th %ile	SD
1. Stature	1535	1640	1745	64	1425	1525	1620	59*
2. Eye height	1430	1505	1640	63	1325	1420	1520	59*
3. Shoulder height	1280	1375	1475	60	1180	1270	1360	60*
4. Elbow height	930	1015	1100	70	875	940	1015	62*
5. Hip height	885	970	1060	59	840	920	985	77*
6. Knuckle height	630	700	770	42	580	655	730	45
7. Fingertip height	555	605	665	39	505	570	675	45*
8. Sitting height	780	835	890	33	725	775	830	32*
9. Sitting eye height	680	730	785	32	625	675	725	30*
10. Sitting shoulder height	520	570	620	30	475	525	575	29*
11. Sitting elbow height	160	200	240	25	150	185	220	20*
12. Thigh thickness	100	120	140	11	70	85	100	9*
13. Buttock–knee length	510	555	620	32	485	535	570	29*
14. Buttock–popliteal length	415	460	510	31	400	445	495	35*
15. Knee height	410	455	500	29	375	420	465	27*
16. Popliteal height	325	370	410	26	290	335	380	27
17. Shoulder breadth (bideltoid)	330	370	400	23	300	330	360	18*
18. Shoulder breadth (biacromial)	295	320	345	16	270	295	320	14
19. Hip breadth	225	250	280	18	210	245	280	22*
20. Chest (bust) depth	145	170	205	20	130	170	210	25*(M)
21. Abdominal depth	140	185	235	29	130	175	220	28
22. Shoulder–elbow length	290	345	400	33	270	315	355	26
23. Elbow fingertip length	415	450	490	34	380	410	450	29*
24. Upper limb length	680	735	790	33	615	665	715	30
25. Shoulder–grip length	615	665	710	29	560	605	650	28
26. Head length	155	180	195	12	150	170	190	20*
27. Head breadth	130	145	155	12	125	135	145	17*
28. Hand length	165	180	195	12	150	165	180	14*
29. Hand breadth	90	100	110	6	80	90	90	5*
30. Foot length	230	250	280	17	210	235	255	13*
31. Foot breadth	90	105	120	11	80	95	110	12*
32. Span	1505	1690	1815	89	1405	1545	1680	99*
33. Elbow span	780	880	940	51	720	795	870	46*
34. Vertical grip reach (standing)	1895	1940	2060	73	1695	1805	1915	68
35. Vertical grip reach (sitting)	1085	1175	1265	55	1010	1090	1170	50
36. Forward grip reach	685	735	785	31	620	670	720	30
Body weight	*39*	*51*	*67*	*8*	*34*	*43*	*56*	*7*

See notes on p. 176.

Table 10.14 Anthropometric estimates for Indian agricultural workers (all dimensions in mm, except for body weight, given in kg).

	Men			
	5th %ile	50th %ile	95th %ile	SD
1. Stature	1540	1620	1700	50*
2. Eye height	1425	1510	1595	52*
3. Shoulder height	1265	1345	1425	49*
4. Elbow height	940	1025	1105	40*
5. Hip height	800	865	930	38
6. Knuckle height	635	685	730	29*
7. Fingertip height	540	585	630	28*
8. Sitting height	795	840	880	25*
9. Sitting eye height	695	740	780	26*
10. Sitting shoulder height	520	555	590	21*
11. Sitting elbow height	170	205	235	20*
12. Thigh thickness	110	135	160	13*
13. Buttock–knee length	520	555	590	21*
14. Buttock–popliteal length	435	465	495	18*
15. Knee height	460	510	560	30*
16. Popliteal height	380	415	450	21*
17. Shoulder breadth (bideltoid)	375	410	440	19*
18. Shoulder breadth (biacromial)	320	355	395	24*
19. Hip breadth	280	310	335	16*
20. Chest depth	145	170	205	20
21. Abdominal depth	140	185	235	33
22. Shoulder–elbow length	325	355	385	19
23. Elbow fingertip length	425	460	490	20*
24. Upper limb length	700	755	810	34
25. Shoulder–grip length	655	710	760	32*
26. Head length	170	180	190	6
27. Head breadth	140	145	150	4
28. Hand length	170	185	195	8*
29. Hand breadth	75	85	90	4*
30. Foot length	235	250	265	10*
31. Foot breadth	90	95	105	4*
32. Span	1595	1705	1810	66*
33. Elbow span	825	880	935	33*
34. Vertical grip reach (standing)	1875	1995	2110	72
35. Vertical grip reach (sitting)	1120	1190	1265	44*
36. Forward grip reach	685	725	765	24
Body weight	*40*	*49*	*59*	*6**

See notes on p. 176.

Table 10.15 Anthropometric estimates for Hong Kong Chinese industrial workers (all dimensions in mm, except for bodyweight, given in kg).

Dimension	Men				Women			
	5th %ile	50th %ile	95th %ile	SD	5th %ile	50th %ile	95th %ile	SD
1. Stature	1585	1680	1775	58	1455	1555	1655	60*
2. Eye height	1470	1555	1640	52	1330	1425	1520	57*
3. Shoulder height	1300	1380	1460	50	1180	1265	1350	51*
4. Elbow height	950	1015	1080	39	870	935	1000	41*
5. Hip height	790	855	920	41	715	785	855	42
6. Knuckle height	685	750	815	40	650	715	780	41
7. Fingertip height	575	640	705	38	540	610	680	44
8. Sitting height	845	900	955	34	780	840	900	37*
9. Sitting eye height	720	780	840	35	660	720	780	35*
10. Sitting shoulder height	555	605	655	31	510	560	610	29*
11. Sitting elbow height	190	240	290	31	165	230	295	38*
12. Thigh thickness	110	135	160	14	105	130	155	14
13. Buttock–knee length	505	550	595	26	470	520	570	30*
14. Buttock–popliteal length	405	450	495	26	385	435	485	29*
15. Knee height	450	495	540	26	410	455	500	27*
16. Popliteal height	365	405	445	25	325	375	425	29*
17. Shoulder breadth (bideltoid)	380	425	470	26	335	385	435	29*
18. Shoulder breadth (biacromial)	335	365	395	19	315	350	385	22
19. Hip breadth	300	335	370	22	295	330	365	21*(M)
20. Chest (bust) depth	155	195	235	25	160	215	270	34
21. Abdominal depth	150	210	270	36	150	215	280	39
22. Shoulder–elbow length	310	340	370	19	290	315	340	16*
23. Elbow–fingertip length	410	445	480	22	360	400	440	24*
24. Upper limb length	680	730	780	30	615	660	705	26
25. Shoulder–grip length	580	620	660	25	525	560	595	22
26. Head length	175	190	205	8	160	175	190	9
27. Head breadth	150	160	170	7	135	150	165	8
28. Hand length	165	180	195	9	150	165	180	9*
29. Hand breadth	70	80	90	5	60	70	80	5*
30. Foot length	235	250	265	10	205	225	245	11*
31. Foot breadth	85	95	105	5	80	85	90	4*
32. Span	1480	1635	1790	95	1350	1480	1610	80*
33. Elbow span	805	885	965	48	690	775	860	51
34. Vertical grip reach (standing)	1835	1970	2105	83	1685	1825	1965	86
35. Vertical grip reach (sitting)	1110	1205	1300	58	855	940	1025	51
36. Forward grip reach	640	705	770	38	580	635	690	32
Body weight	*47*	*60*	*75*	*9*	*39*	*47*	*62*	*7*

See notes on p. 176.

Table 10.16 Anthropometric estimates for Japanese adults (all dimensions in mm, except body weight, given in kg).

Dimension	Men				Women			
	5th %ile	50th %ile	95th %ile	SD	5th %ile	50th %ile	95th %ile	SD
1. Stature	1560	1655	1750	58	1450	1530	1610	48*
2. Eye height	1445	1540	1635	57	1350	1425	1500	47
3. Shoulder height	1250	1340	1430	54	1075	1145	1215	44
4. Elbow height	965	1035	1105	43	895	955	1015	36
5. Hip height	765	830	895	41	700	755	810	33
6. Knuckle height	675	740	805	40	650	705	760	33
7. Fingertip height	565	630	695	38	540	600	660	35
8. Sitting height	850	900	950	31	800	845	890	28
9. Sitting eye height	735	785	835	31	690	735	780	28
10. Sitting shoulder height	545	590	635	28	510	555	600	26
11. Sitting elbow height	220	260	300	23	215	250	285	20
12. Thigh thickness	110	135	160	14	105	130	155	14
13. Buttock–knee length	500	550	600	29	485	530	575	26
14. Buttock–popliteal length	410	470	510	31	405	450	495	26
15. Knee height	450	490	530	23	420	450	480	18
16. Popliteal height	360	400	440	24	325	360	395	21
17. Shoulder breadth (bideltoid)	405	440	475	22	365	395	425	18
18. Shoulder breadth (biacromial)	350	380	410	18	315	340	365	15
19. Hip breadth	280	305	330	14	270	305	340	20
20. Chest (bust) depth	180	205	230	16	175	205	235	18
21. Abdominal depth	185	220	255	22	170	205	240	20
22. Shoulder–elbow length	295	330	365	21	270	300	330	17
23. Elbow–fingertip length	405	440	475	20	370	400	430	17
24. Upper limb length	665	715	765	29	605	645	685	25
25. Shoulder–grip length	565	610	655	26	515	550	585	22
26. Head length	170	185	200	8	160	170	180	7
27. Head breadth	145	155	165	7	140	150	160	6
28. Hand length	165	180	195	10	150	165	180	9
29. Hand breadth	75	85	95	6	65	75	85	5
30. Foot length	230	245	260	10	210	225	240	9
31. Foot breadth	95	105	115	5	90	95	100	4
32. Span	1540	1655	1770	70	1395	1485	1575	56
33. Elbow span	790	870	950	48	715	780	845	41
34. Vertical grip reach (standing)	1805	1940	2075	83	1680	1795	1910	69
35. Vertical grip reach (sitting)	1105	1185	1265	49	1030	1095	1160	41
36. Forward grip reach	630	690	750	37	570	620	670	31
Body weight	*41*	*60*	*74*	*9*	*40*	*51*	*63*	*7*

See notes on p. 176.

Table 10.17 Anthropometric estimates for newborn infants (all dimensions in mm).

Dimension	5th %ile	50th %ile	95th %ile	SD
1. Crown–heel length (1)[a]	465	500	535	20
2. Crown–rump length (8)	330	350	370	11
3. Rump–knee length (13)	105	125	140	10
4. Knee–sole length (15)	120	135	145	7
5. Shoulder breadth (bideltoid) (17)	135	150	160	8
6. Hip breadth (19)	105	120	130	8
7. Shoulder–elbow length (22)	85	95	105	6
8. Elbow–fingertip length (23)	120	135	145	7
9. Head length (26)	115	120	125	4
10. Head breadth (27)	90	95	100	3
11. Hand length (28)	55	60	65	3
12. Hand breadth (29)	30	35	35	2
13. Foot length (30)	65	75	80	4
14. Foot breadth (31)	30	30	35	2

Note: [a] Numbers in parentheses are the equivalent dimensions in adult tables.
See notes on p. 176.

Table 10.18 Anthropometric estimates for infants less than 6 months of age (all dimensions in mm).

Dimension	5th %ile	50th %ile	95th %ile	SD
1. Crown–heel length (1)[a]	510	600	690	54
2. Crown–rump length (8)	360	410	460	31
3. Rump–knee length (13)	105	150	195	28
4. Knee–sole length (15)	125	160	190	19
5. Shoulder breadth (bideltoid) (17)	140	180	215	21
6. Hip breadth (19)	100	140	175	22
7. Shoulder–elbow length (22)	90	115	145	16
8. Elbow–fingertip length (23)	125	160	190	19
9. Head length (26)	120	140	160	11
10. Head breadth (27)	100	110	120	7
11. Hand length (28)	55	70	85	9
12. Hand breadth (29)	30	40	45	4
13. Foot length (30)	70	85	105	12
14. Foot breadth (31)	30	40	50	6

Note: [a] Numbers in parentheses are the equivalent dimensions in adult tables.
See notes on p. 176.

Table 10.19 Anthropometric estimates for infants from 6 months to 1 year (all dimensions in mm).

Dimension	5th %ile	50th %ile	95th %ile	SD
1. Crown–heel length (1)[a]	655	715	775	37
2. Crown–rump length (8)	435	470	505	21
3. Rump–knee length (13)	155	185	215	19
4. Knee–sole length (15)	170	190	210	13
5. Shoulder breadth (bideltoid) (17)	185	210	230	14
6. Hip breadth (19)	140	165	190	15
7. Shoulder–elbow length (22)	120	140	155	11
8. Elbow–fingertip length (23)	170	190	210	13
9. Head length (26)	145	160	170	8
10. Head breadth (27)	115	120	130	5
11. Hand length (28)	75	85	95	6
12. Hand breadth (29)	40	45	50	3
13. Foot length (30)	90	105	120	8
14. Foot breadth (31)	40	45	50	4

Note: [a] Numbers in parentheses are the equivalent dimensions in adult tables.
See notes on p. 176.

Table 10.20 Anthropometric estimates for infants from 1 year to 18 months (all dimensions in mm).

Dimension	5th %ile	50th %ile	95th %ile	SD
1. Crown–heel length (1)[a]	690	745	800	35
2. Crown–rump length (8)	440	475	505	20
3. Rump–knee length (13)	170	195	225	18
4. Knee–sole length (15)	175	195	215	12
5. Shoulder breadth (bideltoid) (17)	185	205	230	14
6. Hip breadth (19)	140	165	185	14
7. Shoulder–elbow length (22)	130	145	160	10
8. Elbow–fingertip length (23)	175	195	215	12
9. Head length (26)	150	160	170	7
10. Head breadth (27)	115	120	130	5
11. Hand length (28)	80	90	100	6
12. Hand breadth (29)	40	45	50	3
13. Foot length (30)	100	115	125	8
14. Foot breadth (31)	40	45	55	4

Note: [a] Numbers in parentheses are the equivalent dimensions in adult tables.
See notes on p. 176.

Table 10.21 Anthropometric estimates for infants from 18 months to 2 years (all dimensions in mm).

Dimension	5th %ile	50th %ile	95th %ile	SD
1. Crown–heel length (1)[a]	780	840	900	36
2. Crown–rump length (8)	490	525	555	20
3. Rump–knee length (13)	200	230	260	18
4. Knee–sole length (15)	210	230	255	13
5. Shoulder breadth (bideltoid) (17)	205	230	250	14
6. Hip breadth (19)	150	175	200	15
7. Shoulder–elbow length (22)	145	165	180	11
8. Elbow–fingertip length (23)	200	220	240	12
9. Head length (26)	160	175	185	7
10. Head breadth (27)	125	130	140	5
11. Hand length (28)	85	95	105	6
12. Hand breadth (29)	45	50	55	3
13. Foot length (30)	115	125	140	8
14. Foot breadth (31)	45	55	60	4

Note: [a] Numbers in parentheses are the equivalent dimensions in adult tables.
See notes on p. 176–7.

Table 10.22 Anthropometric estimates for British 2-year-olds (all dimensions in mm).

Dimension	Boys 5th %ile	50th %ile	95th %ile	SD	Girls 5th %ile	50th %ile	95th %ile	SD
1. Stature	850	930	1010	49	825	890	955	40*
2. Eye height	760	840	920	49	725	805	885	48
3. Shoulder height	675	735	795	37	630	695	760	38
4. Elbow height	495	555	615	36	480	530	580	30
5. Hip height	360	420	480	37	365	415	465	30
6. Knuckle height	340	385	430	26	335	375	415	25
7. Fingertip height	275	315	360	26	270	310	350	25
8. Sitting height	505	545	585	24	485	520	555	21
9. Sitting eye height	410	445	480	20	370	410	450	24
10. Sitting shoulder height	305	340	375	22	275	310	345	20
11. Sitting elbow height	105	140	175	20	105	130	155	15
12. Thigh thickness	65	80	95	10	60	75	90	10
13. Buttock–knee length	245	275	305	19	250	280	310	17
14. Buttock–popliteal length	210	235	260	16	185	245	305	36
15. Knee height	235	270	305	20	230	260	290	17
16. Popliteal height	155	205	255	29	170	205	240	20
17. Shoulder breadth (bideltoid)	215	245	275	17	210	235	260	14
18. Shoulder breadth (biacromial)	190	215	240	15	190	210	230	12
19. Hip breadth	170	190	210	13	165	185	205	11
20. Chest (bust) depth	100	120	140	12	100	115	130	10
21. Abdominal depth	130	145	160	10	135	145	155	7
22. Shoulder–elbow length	160	185	205	13	160	175	190	10
23. Elbow–fingertip length	215	245	275	17	210	235	260	14
24. Upper limb length	365	410	455	28	335	380	425	27
25. Shoulder–grip length	295	340	390	28	270	315	360	27
26. Head length	170	180	190	7	160	165	170	4
27. Head breadth	130	140	150	6	125	130	135	4
28. Hand length	90	105	120	8	90	100	110	6
29. Hand breadth	50	55	60	4	40	45	50	4
30. Foot length	130	145	160	10	130	145	160	9
31. Foot breadth	60	65	70	4	50	55	60	4
32. Span	835	925	1015	54	785	865	945	49
33. Elbow span	435	490	540	31	410	455	505	30
34. Vertical grip reach (standing)	920	1045	1170	77	965	1045	1125	49
35. Vertical grip reach (sitting)	605	675	745	42	550	620	690	42
36. Forward grip reach	340	400	460	35	345	385	425	25

See notes on p. 176–7.

Table 10.23 Anthropometric estimates for British 3-year-olds (all dimensions in mm).

Dimension	Boys				Girls			
	5th %ile	50th %ile	95th %ile	SD	5th %ile	50th %ile	95th %ile	SD
1. Stature	910	990	1070	48	895	970	1045	46*
2. Eye height	810	890	970	48	785	875	965	55
3. Shoulder height	720	780	840	36	690	760	830	43
4. Elbow height	535	595	655	35	520	580	640	35
5. Hip height	400	460	520	35	405	460	515	33
6. Knuckle height	365	410	455	26	360	410	460	29
7. Fingertip height	295	340	380	26	290	340	385	29
8. Sitting height	530	570	610	25	515	555	595	25
9. Sitting eye height	425	465	505	24	400	445	490	28
10. Sitting shoulder height	310	350	390	23	295	335	375	23
11. Sitting elbow height	115	150	185	20	110	140	170	17
12. Thigh thickness	65	85	105	11	60	80	100	12
13. Buttock–knee length	270	300	330	19	270	305	340	20
14. Buttock–popliteal length	225	250	275	16	215	260	305	26
15. Knee height	255	290	325	20	250	285	320	20
16. Popliteal height	195	230	265	21	200	230	260	17
17. Shoulder breadth (bideltoid)	230	255	280	16	225	250	275	15
18. Shoulder breadth (biacromial)	200	225	250	14	205	225	245	13
19. Hip breadth	175	195	215	13	175	195	215	13
20. Chest (bust) depth	105	125	145	12	105	120	140	12
21. Abdominal depth	135	150	165	10	135	150	165	10
22. Shoulder–elbow length	175	195	220	13	175	195	215	12
23. Elbow–fingertip length	235	260	285	16	230	255	280	16
24. Upper limb length	390	435	480	27	365	415	465	31
25. Shoulder–grip length	320	365	410	27	295	345	395	31
26. Head length	170	180	190	7	155	165	175	6
27. Head breadth	130	140	150	6	120	130	140	5
28. Hand length	95	110	125	8	100	110	120	7
29. Hand breadth	50	55	60	4	45	50	55	4
30. Foot length	140	155	170	10	140	155	170	10
31. Foot breadth	60	65	70	4	55	60	65	4
32. Span	890	980	1070	56	850	940	1030	56
33. Elbow span	465	515	570	32	440	495	555	34
34. Vertical grip reach (standing)	1005	1130	1255	75	1025	1125	1225	61
35. Vertical grip reach (sitting)	640	705	775	41	605	675	740	41
36. Forward grip reach	360	420	480	35	360	415	470	33

See notes on pp. 176–7.

Table 10.24 Anthropometric estimates for British 4-year-olds (all dimensions in mm).

Dimension	Boys				Girls			
	5th %ile	50th %ile	95th %ile	SD	5th %ile	50th %ile	95th %ile	SD
1. Stature	975	1050	1125	47	965	1050	1135	52*
2. Eye height	865	940	1015	47	845	945	1045	62
3. Shoulder height	765	825	885	35	745	825	905	48
4. Elbow height	580	635	690	34	565	630	695	40
5. Hip height	445	500	555	33	445	505	565	36
6. Knuckle height	390	435	480	26	390	445	500	33
7. Fingertip height	315	360	405	26	315	365	420	33
8. Sitting height	550	595	640	26	540	590	640	29
9. Sitting eye height	440	485	530	28	425	480	535	32
10. Sitting shoulder height	320	360	400	24	315	360	405	26
11. Sitting elbow height	125	160	195	20	120	150	180	19
12. Thigh thickness	70	90	110	12	60	85	110	14
13. Buttock–knee length	295	325	355	19	290	330	370	23
14. Buttock–popliteal length	240	265	290	16	250	275	300	16
15. Knee height	275	310	345	20	270	310	350	23
16. Popliteal height	235	255	275	13	230	255	280	14
17. Shoulder breadth (bideltoid)	240	265	290	15	240	265	290	16
18. Shoulder breadth (biacromial)	215	235	255	13	215	240	265	14
19. Hip breadth	180	200	220	13	180	205	230	15
20. Chest (bust) depth	110	130	150	12	110	130	150	13
21. Abdominal depth	140	155	170	10	135	155	175	13
22. Shoulder–elbow length	190	210	230	13	185	210	235	14
23. Elbow–fingertip length	250	275	300	15	245	275	305	18
24. Upper limb length	415	460	505	26	390	450	510	35
25. Shoulder–grip length	340	385	430	26	315	370	430	35
26. Head length	170	180	190	7	150	165	180	8
27. Head breadth	130	140	150	6	125	135	145	6
28. Hand length	100	115	130	8	105	120	135	8
29. Hand breadth	50	55	60	4	50	55	60	4
30. Foot length	150	165	180	10	145	165	185	11
31. Foot breadth	60	65	70	4	60	65	70	4
32. Span	940	1035	1130	58	910	1015	1120	63
33. Elbow span	490	545	600	33	475	535	600	38
34. Vertical grip reach (standing)	1095	1215	1335	73	1085	1205	1325	73
35. Vertical grip reach (sitting)	670	735	805	40	660	725	795	40
36. Forward grip reach	380	440	500	35	380	445	510	41

See notes on pp. 176–7.

Table 10.25 Anthropometric estimates for British 5-year-olds (all dimensions in mm).

Dimension	Boys				Girls			
	5th %ile	50th %ile	95th %ile	SD	5th %ile	50th %ile	95th %ile	SD
1. Stature	1025	1110	1195	52	1015	1100	1185	53*
2. Eye height	910	995	1080	53	885	990	1095	64
3. Shoulder height	810	875	940	39	785	865	945	48
4. Elbow height	605	670	735	39	595	660	725	41
5. Hip height	490	550	610	36	490	540	590	31
6. Knuckle height	405	455	505	29	410	465	520	34
7. Fingertip height	325	375	420	29	330	385	445	34
8. Sitting height	575	620	665	28	560	610	660	29
9. Sitting eye height	455	505	555	29	450	500	550	31
10. Sitting shoulder height	340	380	420	25	325	370	415	26
11. Sitting elbow height	130	165	200	22	125	155	185	19
12. Thigh thickness	75	90	105	10	70	90	110	12
13. Buttock–knee length	310	345	380	21	310	350	390	23
14. Buttock–popliteal length	250	280	310	17	265	295	325	19
15. Knee height	300	335	370	22	295	330	365	21
16. Popliteal height	240	270	300	18	245	270	295	16
17. Shoulder breadth (bideltoid)	245	275	305	17	245	270	295	16
18. Shoulder breadth (biacromial)	225	250	275	15	230	250	270	12
19. Hip breadth	185	210	235	15	185	210	235	16
20. Chest (bust) depth	110	135	160	14	110	135	155	14
21. Abdominal depth	135	155	175	12	135	160	185	14
22. Shoulder–elbow length	205	225	250	14	200	220	245	13
23. Elbow–fingertip length	265	295	325	17	260	290	320	17
24. Upper limb length	435	485	535	29	410	470	530	35
25. Shoulder–grip length	355	405	450	29	335	390	450	35
26. Head length	165	180	195	8	150	165	180	8
27. Head breadth	130	140	150	5	120	130	140	5
28. Hand length	110	125	140	9	105	120	135	8
29. Hand breadth	55	60	65	4	50	55	60	4
30. Foot length	155	175	195	11	155	170	185	10
31. Foot breadth	60	70	80	5	60	65	70	4
32. Span	995	1095	1195	60	955	1060	1165	64
33. Elbow span	520	575	635	34	495	560	625	39
34. Vertical grip reach (standing)	1180	1305	1430	77	1170	1290	1410	72
35. Vertical grip reach (sitting)	700	775	850	45	685	755	830	45
36. Forward grip reach	420	470	520	30	400	460	520	36

See notes on pp. 176–7.

Table 10.26 Anthropometric estimates for British 6-year-olds (all dimensions in mm).

	Boys				Girls			
Dimension	5th %ile	50th %ile	95th %ile	SD	5th %ile	50th %ile	95th %ile	SD
1. Stature	1070	1170	1270	60	1070	1160	1250	56*
2. Eye height	950	1050	1150	60	935	1045	1155	67
3. Shoulder height	845	920	995	45	825	910	995	52
4. Elbow height	635	705	775	44	625	695	765	43
5. Hip height	520	595	670	45	420	475	530	32
6. Knuckle height	425	480	535	33	430	490	550	36
7. Fingertip height	340	395	450	33	350	410	470	36
8. Sitting height	585	640	695	32	585	635	685	31
9. Sitting eye height	475	525	575	31	470	525	580	32
10. Sitting shoulder height	340	390	440	29	335	380	425	28
11. Sitting elbow height	130	170	210	25	125	160	195	21
12. Thigh thickness	75	95	115	13	75	95	115	11
13. Buttock–knee length	330	370	410	25	330	370	410	25
14. Buttock–popliteal length	270	305	340	21	275	310	345	20
15. Knee height	320	360	400	25	320	355	390	21
16. Popliteal height	260	295	330	22	265	290	315	16
17. Shoulder breadth (bideltoid)	245	285	325	23	250	285	320	20
18. Shoulder breadth (biacromial)	235	265	295	18	240	260	280	13
19. Hip breadth	180	215	250	21	190	220	250	19
20. Chest (bust) depth	110	140	170	19	110	140	170	18
21. Abdominal depth	135	160	185	16	135	165	195	18
22. Shoulder–elbow length	215	240	265	16	215	235	255	13
23. Elbow–fingertip length	275	310	345	21	275	305	335	18
24. Upper limb length	455	510	565	34	430	495	560	38
25. Shoulder–grip length	370	425	480	34	350	415	475	38
26. Head length	165	180	195	9	160	170	180	7
27. Head breadth	130	140	150	6	125	135	145	6
28. Hand length	115	130	145	10	110	125	140	8
29. Hand breadth	50	60	70	5	55	60	65	4
30. Foot length	165	185	205	13	160	180	200	11
31. Foot breadth	65	75	85	6	60	70	80	5
32. Span	1045	1160	1275	70	1010	1120	1230	68
33. Elbow span	545	610	675	40	525	590	660	41
34. Vertical grip reach (standing)	1235	1390	1545	93	1255	1380	1505	76
35. Vertical grip reach (sitting)	720	805	890	52	705	790	875	52
36. Forward grip reach	435	495	555	35	435	485	535	31

See notes on pp. 176–7.

Table 10.27 Anthropometric estimates for British 7-year-olds (all dimensions in mm).

	Boys				Girls			
Dimension	5th %ile	50th %ile	95th %ile	SD	5th %ile	50th %ile	95th %ile	SD
1. Stature	1140	1230	1320	56	1125	1220	1315	59*
2. Eye height	1020	1115	1210	57	995	1105	1215	66
3. Shoulder height	885	975	1065	54	870	960	1050	54
4. Elbow height	680	745	810	40	665	735	805	42
5. Hip height	570	635	700	39	555	615	675	35
6. Knuckle height	460	510	560	31	465	525	585	37
7. Fingertip height	370	420	475	31	375	435	500	37
8. Sitting height	615	665	715	30	610	660	710	31
9. Sitting eye height	505	550	595	28	500	555	610	32
10. Sitting shoulder height	360	405	450	27	350	395	440	26
11. Sitting elbow height	140	175	210	20	140	170	200	19
12. Thigh thickness	85	105	125	13	85	105	125	13
13. Buttock–knee length	355	395	435	24	355	400	445	26
14. Buttock–popliteal length	280	325	370	27	290	335	380	27
15. Knee height	340	380	420	25	335	375	415	23
16. Popliteal height	285	315	345	19	275	310	345	21
17. Shoulder breadth (bideltoid)	265	300	335	22	255	295	335	24
18. Shoulder breadth (biacromial)	250	275	300	15	245	270	295	15
19. Hip breadth	190	225	260	21	195	235	275	23
20. Chest (bust) depth	110	145	180	20	110	145	180	21
21. Abdominal depth	135	165	195	19	130	170	210	23
22. Shoulder–elbow length	230	255	280	15	225	250	275	15
23. Elbow–fingertip length	295	325	355	19	290	320	350	18
24. Upper limb length	485	540	595	32	470	525	580	34
25. Shoulder–grip length	400	450	505	32	380	435	495	34
26. Head length	170	185	200	8	160	170	180	6
27. Head breadth	130	140	150	5	125	135	145	5
28. Hand length	120	135	150	9	120	135	150	8
29. Hand breadth	60	65	70	4	55	60	65	4
30. Foot length	175	195	215	11	170	190	210	12
31. Foot breadth	65	75	85	5	65	75	85	5
32. Span	1125	1230	1335	64	1095	1195	1295	62
33. Elbow span	590	650	710	36	570	630	695	38
34. Vertical grip reach (standing)	1350	1475	1600	76	1325	1455	1585	79
35. Vertical grip reach (sitting)	770	850	925	48	745	825	905	48
36. Forward grip reach	470	520	570	31	455	505	555	29

See notes on pp. 176–7.

Table 10.28 Anthropometric estimates for British 8-year-olds (all dimensions in mm).

Dimension	Boys 5th %ile	50th %ile	95th %ile	SD	Girls 5th %ile	50th %ile	95th %ile	SD
1. Stature	1180	1280	1380	60	1185	1280	1375	59*
2. Eye height	1070	1165	1260	59	1070	1165	1260	58
3. Shoulder height	930	1020	1110	54	930	1015	1100	53
4. Elbow height	705	780	855	45	705	775	845	42
5. Hip height	605	665	725	35	585	650	715	38
6. Knuckle height	480	535	590	32	495	555	615	37
7. Fingertip height	390	445	495	32	405	465	525	37
8. Sitting height	630	680	730	31	640	685	730	28
9. Sitting eye height	520	570	620	31	525	580	635	32
10. Sitting shoulder height	380	425	470	27	370	410	450	25
11. Sitting elbow height	145	180	215	21	145	175	205	19
12. Thigh thickness	85	110	135	14	90	110	130	13
13. Buttock–knee length	375	415	455	25	375	420	465	26
14. Buttock–popliteal length	305	340	375	22	310	355	400	27
15. Knee height	360	400	440	25	355	395	435	24
16. Popliteal height	295	325	355	18	295	330	365	20
17. Shoulder breadth (bideltoid)	275	310	345	21	270	310	350	24
18. Shoulder breadth (biacromial)	265	285	305	13	255	280	305	16
19. Hip breadth	200	235	270	20	205	245	285	23
20. Chest (bust) depth	115	150	185	20	120	150	180	20
21. Abdominal depth	135	170	205	20	140	180	220	24
22. Shoulder–elbow length	240	265	290	15	240	260	285	14
23. Elbow–fingertip length	310	340	370	19	305	335	365	19
24. Upper limb length	515	565	615	30	495	555	615	35
25. Shoulder–grip length	425	475	525	30	405	465	520	35
26. Head length	170	185	200	8	165	175	185	5
27. Head breadth	130	140	150	5	125	135	145	5
28. Hand length	125	140	155	9	125	140	155	8
29. Hand breadth	60	65	70	4	60	65	70	4
30. Foot length	180	200	220	12	180	200	220	12
31. Foot breadth	70	80	90	5	65	75	85	5
32. Span	1165	1280	1395	69	1150	1250	1350	60
33. Elbow span	610	675	740	39	600	660	720	36
34. Vertical grip reach (standing)	1425	1550	1675	75	1405	1535	1665	78
35. Vertical grip reach (sitting)	805	890	975	52	785	870	955	52
36. Forward grip reach	475	535	595	35	475	530	585	34

See notes on pp. 176–7.

Table 10.29 Anthropometric estimates for British 9-year-olds (all dimensions in mm).

	Boys				Girls			
Dimension	5th %ile	50th %ile	95th %ile	SD	5th %ile	50th %ile	95th %ile	SD
1. Stature	1225	1330	1435	63	1220	1330	1440	68*
2. Eye height	1005	1110	1215	64	1105	1215	1325	67
3. Shoulder height	965	1065	1165	60	955	1060	1165	63
4. Elbow height	740	820	900	50	720	815	910	57
5. Hip height	635	700	765	40	610	690	770	48
6. Knuckle height	505	565	625	36	530	590	650	37
7. Fingertip height	410	470	530	36	435	495	555	37
8. Sitting height	650	700	750	31	645	700	755	33
9. Sitting eye height	530	585	640	33	540	595	650	33
10. Sitting shoulder height	390	440	490	29	385	430	475	28
11. Sitting elbow height	150	190	230	24	140	180	220	25
12. Thigh thickness	90	115	140	15	90	115	140	15
13. Buttock–knee length	395	440	485	26	395	445	495	30
14. Buttock–popliteal length	325	365	405	25	330	380	430	31
15. Knee height	375	420	465	27	375	420	465	27
16. Popliteal height	300	340	380	23	300	340	380	24
17. Shoulder breadth (bideltoid)	280	320	360	23	285	320	355	20
18. Shoulder breadth (biacromial)	270	295	320	15	265	295	325	19
19. Hip breadth	205	245	285	24	210	255	300	27
20. Chest (bust) depth	120	155	190	22	115	155	195	24
21. Abdominal depth	145	180	215	21	140	185	230	26
22. Shoulder–elbow length	250	275	305	16	245	275	300	17
23. Elbow–fingertip length	320	355	390	21	310	350	390	23
24. Upper limb length	530	585	640	33	500	575	650	45
25. Shoulder–grip length	435	490	545	33	405	480	555	45
26. Head length	170	185	200	8	165	175	185	7
27. Head breadth	135	145	155	5	125	135	145	6
28. Hand length	130	145	160	9	130	145	160	10
29. Hand breadth	60	65	70	4	60	65	70	4
30. Foot length	185	210	235	14	185	210	235	14
31. Foot breadth	70	80	90	5	70	80	90	6
32. Span	1200	1330	1460	78	1180	1300	1420	74
33. Elbow span	630	700	775	44	615	685	760	45
34. Vertical grip reach (standing)	1475	1610	1745	83	1460	1615	1770	94
35. Vertical grip reach (sitting)	830	920	1010	54	815	905	995	54
36. Forward grip reach	495	555	615	36	485	555	625	42

See notes on pp. 176–7.

Table 10.30 Anthropometric estimates for British 10-year-olds (all dimensions in mm).

Dimension	Boys				Girls			
	5th %ile	50th %ile	95th %ile	SD	5th %ile	50th %ile	95th %ile	SD
1. Stature	1290	1390	1490	61	1270	1390	1510	72*
2. Eye height	1180	1275	1370	58	1155	1275	1395	72
3. Shoulder height	1025	1120	1215	57	1015	1120	1225	65
4. Elbow height	770	860	950	55	765	860	955	57
5. Hip height	660	735	810	46	650	730	810	50
6. Knuckle height	540	595	650	33	555	615	675	36
7. Fingertip height	445	500	550	33	460	520	575	36
8. Sitting height	670	725	780	32	665	725	785	36
9. Sitting eye height	550	600	650	29	555	615	675	35
10. Sitting shoulder height	410	455	500	28	400	450	500	30
11. Sitting elbow height	160	195	230	21	150	190	230	25
12. Thigh thickness	100	120	140	13	95	120	145	16
13. Buttock–knee length	415	460	505	27	415	470	525	32
14. Buttock–popliteal length	340	380	420	25	350	400	450	29
15. Knee height	395	440	485	26	395	440	485	28
16. Popliteal height	330	360	390	19	325	365	405	25
17. Shoulder breadth (bideltoid)	290	335	380	27	280	330	380	31
18. Shoulder breadth (biacromial)	275	305	335	18	275	305	335	19
19. Hip breadth	215	260	305	28	215	265	315	30
20. Chest (bust) depth	120	165	210	26	115	165	215	31
21. Abdominal depth	145	185	225	25	145	190	235	27
22. Shoulder–elbow length	265	290	315	16	260	290	320	18
23. Elbow–fingertip length	335	370	405	22	330	370	410	25
24. Upper limb length	540	610	680	42	520	590	660	44
25. Shoulder–grip length	445	515	580	42	420	495	565	44
26. Head length	170	185	200	8	160	170	180	7
27. Head breadth	135	145	155	5	125	135	145	5
28. Hand length	135	150	165	9	135	150	165	10
29. Hand breadth	65	70	75	4	60	70	80	5
30. Foot length	195	220	245	14	190	215	240	14
31. Foot breadth	70	85	95	5	70	80	90	7
32. Span	1275	1395	1515	73	1240	1365	1490	77
33. Elbow span	665	735	805	41	645	720	800	47
34. Vertical grip reach (standing)	1540	1680	1820	86	1540	1705	1870	101
35. Vertical grip reach (sitting)	870	955	1045	52	850	935	1020	52
36. Forward grip reach	525	580	635	33	520	585	650	40

See notes on pp. 176–7.

Table 10.31 Anthropometric estimates for British 11-year-olds (all dimensions in mm).

	Boys				Girls			
Dimension	5th %ile	50th %ile	95th %ile	SD	5th %ile	50th %ile	95th %ile	SD
1. Stature	1325	1430	1535	65	1310	1440	1570	79*
2. Eye height	1215	1315	1415	62	1195	1325	1455	78
3. Shoulder height	1060	1160	1260	60	1050	1165	1280	69
4. Elbow height	795	890	985	57	800	890	980	56
5. Hip height	685	765	845	50	670	750	830	48
6. Knuckle height	560	620	680	35	575	645	715	42
7. Fingertip height	460	520	575	35	475	545	615	42
8. Sitting height	685	740	795	34	680	745	810	41
9. Sitting eye height	575	620	665	28	570	635	700	39
10. Sitting shoulder height	425	470	515	26	415	470	525	33
11. Sitting elbow height	160	200	240	24	155	200	245	26
12. Thigh thickness	100	120	140	11	100	125	150	16
13. Buttock–knee length	435	480	525	28	430	490	550	37
14. Buttock–popliteal length	345	395	445	30	365	410	455	26
15. Knee height	420	460	500	25	405	455	505	30
16. Popliteal height	330	375	420	26	335	375	415	24
17. Shoulder breadth (bideltoid)	300	345	390	26	285	340	395	34
18. Shoulder breadth (biacromial)	280	315	350	21	280	315	350	21
19. Hip breadth	220	265	310	27	225	280	335	34
20. Chest (bust) depth	130	170	210	24	115	175	240	38
21. Abdominal depth	150	190	230	23	145	195	245	29
22. Shoulder–elbow length	270	300	325	16	265	300	330	20
23. Elbow–fingertip length	350	385	420	22	340	385	430	28
24. Upper limb length	560	630	700	43	555	630	705	46
25. Shoulder–grip length	460	530	600	43	455	530	605	46
26. Head length	170	185	200	8	155	170	185	8
27. Head breadth	135	145	155	5	125	135	145	5
28. Hand length	140	155	170	10	135	155	175	11
29. Hand breadth	60	70	80	5	60	70	80	5
30. Foot length	205	225	245	13	195	220	245	14
31. Foot breadth	75	85	95	7	75	85	95	7
32. Span	1310	1440	1570	78	1270	1415	1560	87
33. Elbow span	685	760	830	44	660	750	835	53
34. Vertical grip reach (standing)	1575	1740	1905	100	1575	1760	1945	111
35. Vertical grip reach (sitting)	895	990	1080	56	900	990	1085	56
36. Forward grip reach	535	595	655	37	530	600	670	42

See notes on pp. 176–7.

Table 10.32 Anthropometric estimates for British 12-year-olds (all dimensions in mm).

Dimension	Boys				Girls			
	5th %ile	50th %ile	95th %ile	SD	5th %ile	50th %ile	95th %ile	SD
1. Stature	1360	1490	1620	78	1370	1500	1630	79*
2. Eye height	1245	1375	1505	78	1255	1385	1515	80
3. Shoulder height	1095	1215	1335	72	1100	1215	1330	69
4. Elbow height	840	930	1020	55	840	940	1040	60
5. Hip height	720	805	890	53	705	780	855	47
6. Knuckle height	580	645	710	40	590	665	740	46
7. Fingertip height	470	540	605	40	480	560	635	46
8. Sitting height	700	765	830	39	700	775	850	45
9. Sitting eye height	590	650	710	37	600	665	730	40
10. Sitting shoulder height	440	490	540	30	435	490	545	32
11. Sitting elbow height	160	205	250	27	155	205	255	31
12. Thigh thickness	105	125	145	13	100	130	160	17
13. Buttock–knee length	445	500	555	32	450	510	570	36
14. Buttock–popliteal length	375	415	455	23	380	435	490	33
15. Knee height	430	480	530	30	420	470	520	29
16. Popliteal height	350	390	430	23	345	385	425	24
17. Shoulder breadth (bideltoid)	315	355	395	25	305	355	405	29
18. Shoulder breadth (biacromial)	290	325	360	21	290	325	360	21
19. Hip breadth	230	275	320	26	235	295	355	35
20. Chest (bust) depth	135	175	215	24	135	190	240	33
21. Abdominal depth	165	200	235	22	155	200	245	27
22. Shoulder–elbow length	280	310	340	18	280	315	345	20
23. Elbow–fingertip length	360	400	440	25	355	400	445	27
24. Upper limb length	600	665	730	41	575	660	745	52
25. Shoulder–grip length	490	560	625	41	465	555	640	52
26. Head length	170	185	200	8	165	175	185	7
27. Head breadth	135	145	155	5	130	140	150	6
28. Hand length	150	165	180	10	145	165	185	11
29. Hand breadth	65	75	85	5	60	70	80	5
30. Foot length	215	235	255	13	205	230	255	14
31. Foot breadth	80	90	100	7	75	85	95	7
32. Span	1355	1510	1665	93	1320	1480	1640	96
33. Elbow span	710	795	885	53	685	780	880	58
34. Vertical grip reach (standing)	1655	1835	2015	110	1650	1835	2020	112
35. Vertical grip reach (sitting)	925	1035	1145	67	925	1035	1145	67
36. Forward grip reach	550	620	690	42	550	625	700	45

See notes on pp. 176–7.

Table 10.33   Anthropometric estimates for British 13-year-olds (all dimensions in mm).

Dimension	Boys				Girls			
	5th %ile	50th %ile	95th %ile	SD	5th %ile	50th %ile	95th %ile	SD
1. Stature	1400	1550	1700	91	1430	1550	1670	73*
2. Eye height	1285	1435	1585	90	1315	1435	1555	74
3. Shoulder height	1130	1265	1400	81	1145	1255	1365	68
4. Elbow height	870	970	1070	61	875	970	1065	57
5. Hip height	740	835	930	57	725	805	885	50
6. Knuckle height	600	670	740	43	605	675	745	43
7. Fingertip height	490	560	630	43	495	565	635	43
8. Sitting height	710	790	870	49	740	805	870	41
9. Sitting eye height	605	680	755	47	630	695	760	39
10. Sitting shoulder height	450	510	570	37	455	510	565	34
11. Sitting elbow height	165	210	255	28	155	210	265	34
12. Thigh thickness	105	130	155	15	110	135	160	15
13. Buttock–knee length	465	525	585	35	480	530	580	31
14. Buttock–popliteal length	375	435	495	35	400	445	490	27
15. Knee height	440	500	560	35	440	485	530	27
16. Popliteal height	355	405	455	30	350	390	430	25
17. Shoulder breadth (bideltoid)	325	375	425	29	325	370	415	26
18. Shoulder breadth (biacromial)	295	335	375	24	300	335	370	21
19. Hip breadth	245	290	335	28	265	315	365	30
20. Chest (bust) depth	135	185	235	29	150	200	245	29
21. Abdominal depth	165	205	245	24	170	210	250	24
22. Shoulder–elbow length	290	325	360	22	295	325	355	18
23. Elbow–fingertip length	370	420	470	29	375	410	445	22
24. Upper limb length	620	695	770	47	605	680	755	47
25. Shoulder–grip length	505	585	660	47	490	570	645	47
26. Head length	175	190	205	8	165	175	185	6
27. Head breadth	140	150	160	5	130	140	150	5
28. Hand length	150	170	190	12	155	170	185	10
29. Hand breadth	70	80	90	6	70	75	80	4
30. Foot length	220	245	270	16	210	230	250	13
31. Foot breadth	80	90	100	7	80	90	100	6
32. Span	1400	1580	1760	110	1385	1540	1695	93
33. Elbow span	730	835	935	62	720	815	905	56
34. Vertical grip reach (standing)	1720	1905	2090	112	1700	1890	2080	114
35. Vertical grip reach (sitting)	955	1080	1210	78	945	1070	1200	78
36. Forward grip reach	575	655	735	48	575	640	705	41

See notes on pp. 176–7.

Table 10.34   Anthropometric estimates for British 14-year-olds (all dimensions in mm).

Dimension	Boys				Girls			
	5th %ile	50th %ile	95th %ile	SD	5th %ile	50th %ile	95th %ile	SD
1. Stature	1480	1630	1780	90	1480	1590	1700	66*
2. Eye height	1360	1510	1660	91	1365	1475	1585	66
3. Shoulder height	1205	1335	1465	80	1190	1295	1400	64
4. Elbow height	915	1015	1115	60	900	985	1070	53
5. Hip height	795	870	945	46	735	810	885	45
6. Knuckle height	630	700	770	44	640	705	770	40
7. Fingertip height	510	585	655	44	530	595	660	40
8. Sitting height	750	835	920	52	770	830	890	36
9. Sitting eye height	640	720	800	48	660	720	780	37
10. Sitting shoulder height	470	535	600	39	470	525	580	32
11. Sitting elbow height	165	215	265	30	165	220	275	33
12. Thigh thickness	115	140	165	16	115	140	165	14
13. Buttock–knee length	495	550	605	34	495	545	595	29
14. Buttock–popliteal length	405	460	515	33	415	455	495	25
15. Knee height	465	520	575	33	450	495	540	27
16. Popliteal height	380	425	470	28	355	395	435	25
17. Shoulder breadth (bideltoid)	345	395	445	29	345	385	425	25
18. Shoulder breadth (biacromial)	320	355	390	22	315	345	375	19
19. Hip breadth	260	305	350	28	285	330	375	26
20. Chest (bust) depth	145	195	245	29	165	210	255	27
21. Abdominal depth	175	215	255	23	175	215	255	23
22. Shoulder–elbow length	310	345	380	21	305	335	360	17
23. Elbow–fingertip length	400	445	490	28	385	420	455	21
24. Upper limb length	660	735	810	47	640	700	760	36
25. Shoulder–grip length	540	620	695	47	530	590	650	36
26. Head length	180	190	200	7	165	175	185	6
27. Head breadth	140	150	160	6	130	140	150	5
28. Hand length	160	180	200	11	155	170	185	9
29. Hand breadth	75	85	95	6	70	75	80	4
30. Foot length	230	255	280	15	215	235	255	12
31. Foot breadth	85	95	105	7	80	90	100	6
32. Span	1480	1670	1860	114	1450	1580	1710	79
33. Elbow span	775	880	985	65	755	835	915	48
34. Vertical grip reach (standing)	1825	1990	2155	101	1765	1930	2095	101
35. Vertical grip reach (sitting)	1015	1140	1270	77	980	1105	1235	77
36. Forward grip reach	615	680	745	41	595	655	715	36

See notes on pp. 176–7.

Table 10.35 Anthropometric estimates for British 15-year-olds (all dimensions in mm).

Dimension	Boys				Girls			
	5th %ile	50th %ile	95th %ile	SD	5th %ile	50th %ile	95th %ile	SD
1. Stature	1555	1690	1825	83	1510	1610	1710	62*
2. Eye height	1430	1570	1710	84	1395	1495	1595	62
3. Shoulder height	1265	1385	1505	73	1215	1310	1405	58
4. Elbow height	965	1055	1145	56	915	995	1075	48
5. Hip height	825	895	965	44	745	815	885	42
6. Knuckle height	650	725	800	45	650	715	780	39
7. Fingertip height	530	605	680	45	540	605	670	39
8. Sitting height	785	870	955	51	790	845	900	33
9. Sitting eye height	680	755	830	47	680	735	790	33
10. Sitting shoulder height	495	555	615	36	490	535	580	27
11. Sitting elbow height	170	225	280	34	180	225	270	28
12. Thigh thickness	115	140	165	15	115	140	165	14
13. Buttock–knee length	515	570	625	32	505	550	595	27
14. Buttock–popliteal length	425	480	535	32	435	470	505	22
15. Knee height	485	535	585	31	450	495	540	26
16. Popliteal height	385	430	475	28	360	400	440	25
17. Shoulder breadth (bideltoid)	354	415	465	30	350	390	430	23
18. Shoulder breadth (biacromial)	330	370	410	23	320	350	380	18
19. Hip breadth	275	320	365	26	295	335	375	25
20. Chest (bust) depth	155	205	255	30	175	215	260	26
21. Abdominal depth	180	220	260	24	185	220	255	22
22. Shoulder–elbow length	325	355	385	19	305	335	365	17
23. Elbow–fingertip length	420	460	500	25	395	425	455	17
24. Upper limb length	695	770	845	47	650	705	760	33
25. Shoulder–grip length	570	650	725	47	540	595	650	33
26. Head length	185	195	205	7	170	180	190	7
27. Head breadth	145	155	165	6	130	140	150	5
28. Hand length	170	185	200	10	155	170	185	9
29. Hand breadth	75	85	95	5	70	75	80	4
30. Foot length	240	260	280	13	215	235	255	13
31. Foot breadth	85	95	105	7	80	90	100	5
32. Span	1560	1740	1920	109	1490	1600	1710	67
33. Elbow span	815	915	1020	62	780	845	910	41
34. Vertical grip reach (standing)	1900	2060	2220	97	1810	1960	2110	91
35. Vertical grip reach (sitting)	1075	1190	1310	71	1005	1120	1240	71
36. Forward grip reach	635	700	765	40	600	665	730	39

See notes on pp. 176–7.

Table 10.36   Anthropometric estimates for British 16-year-olds (all dimensions in mm).

	Men				Women			
Dimension	5th %ile	50th %ile	95th %ile	SD	5th %ile	50th %ile	95th %ile	SD
1. Stature	1620	1730	1840	68	1520	1620	1720	61*
2. Eye height	1500	1610	1720	67	1410	1510	1610	60
3. Shoulder height	1315	1415	1515	62	1225	1315	1405	55
4. Elbow height	995	1075	1155	49	930	1005	1080	45
5. Hip height	830	910	990	49	755	820	885	40
6. Knuckle height	675	740	805	40	660	720	780	36
7. Fingertip height	555	620	685	40	545	605	665	36
8. Sitting height	830	895	960	39	800	855	910	33
9. Sitting eye height	725	785	845	35	685	740	795	32
10. Sitting shoulder height	520	570	620	29	500	545	590	27
11. Sitting elbow height	190	235	280	28	185	230	275	26
12. Thigh thickness	125	150	175	15	120	145	170	14
13. Buttock–knee length	530	580	630	29	510	555	600	27
14. Buttock–popliteal length	435	490	545	32	435	480	525	26
15. Knee height	500	545	590	27	450	495	540	26
16. Popliteal height	395	440	485	28	365	405	445	25
17. Shoulder breadth (bideltoid)	380	430	480	29	360	395	430	21
18. Shoulder breadth (biacromial)	340	380	420	23	330	355	380	16
19. Hip breadth	290	330	370	23	305	345	385	25
20. Chest (bust) depth	165	215	265	29	180	225	265	25
21. Abdominal depth	185	225	265	24	185	220	255	21
22. Shoulder–elbow length	335	365	395	18	310	335	365	17
23. Elbow–fingertip length	435	470	505	22	395	425	455	17
24. Upper limb length	725	790	855	40	660	710	760	29
25. Shoulder–grip length	605	670	735	40	550	595	645	29
26. Head length	185	195	205	7	165	180	195	8
27. Head breadth	145	155	165	5	135	145	155	5
28. Hand length	170	185	200	9	160	175	190	9
29. Hand breadth	80	85	90	4	70	75	80	4
30. Foot length	240	260	280	12	220	240	260	12
31. Foot breadth	90	100	110	6	80	90	100	5
32. Span	1640	1785	1930	88	1500	1610	1720	67
33. Elbow span	860	940	1025	50	785	850	920	41
34. Vertical grip reach (standing)	1945	2100	2255	93	1820	1965	2110	88
35. Vertical grip reach (sitting)	1130	1225	1320	58	1035	1135	1230	58
36. Forward grip reach	650	720	790	42	605	670	735	41

See notes on pp. 176–7.

Table 10.37 Anthropometric estimates for British 17-year-olds (all dimensions in mm).

Dimension	Boys				Girls			
	5th %ile	50th %ile	95th %ile	SD	5th %ile	50th %ile	95th %ile	SD
1. Stature	1640	1750	1860	66	1520	1620	1720	61*
2. Eye height	1530	1635	1740	65	1420	1515	1610	58
3. Shoulder height	1335	1435	1535	62	1235	1320	1405	52
4. Elbow height	1010	1090	1170	50	935	1005	1075	43
5. Hip height	845	925	1005	50	755	820	885	39
6. Knuckle height	690	755	820	41	670	725	780	33
7. Fingertip height	565	630	700	41	555	610	665	33
8. Sitting height	850	910	970	35	800	855	910	33
9. Sitting eye height	745	795	845	31	690	740	790	30
10. Sitting shoulder height	535	585	635	29	515	555	565	25
11. Sitting elbow height	195	240	285	28	190	230	270	25
12. Thigh thickness	125	155	185	17	120	145	170	16
13. Buttock–knee length	535	585	635	30	515	560	605	28
14. Buttock–popliteal length	445	495	545	30	435	480	525	27
15. Knee height	505	550	595	27	455	500	545	26
16. Popliteal height	405	445	485	25	365	405	445	25
17. Shoulder breadth (bideltoid)	400	445	490	28	360	395	430	21
18. Shoulder breadth (biacromial)	350	385	420	21	335	360	385	16
19. Hip breadth	295	335	375	24	300	345	390	28
20. Chest (bust) depth	180	225	270	27	190	230	270	24
21. Abdominal depth	195	235	275	25	185	220	255	21
22. Shoulder–elbow length	335	365	395	19	305	335	365	18
23. Elbow–fingertip length	440	475	510	21	395	425	455	17
24. Upper limb length	730	790	850	36	660	710	760	29
25. Shoulder–grip length	605	665	725	36	550	595	645	29
26. Head length	185	200	215	8	165	180	195	8
27. Head breadth	145	155	165	6	135	145	155	5
28. Hand length	175	190	205	9	160	175	190	9
29. Hand breadth	80	90	100	5	70	75	80	4
30. Foot length	240	265	290	15	220	240	260	12
31. Foot breadth	90	100	110	5	80	90	100	5
32. Span	1660	1795	1930	81	1510	1615	1720	64
33. Elbow span	870	945	1020	46	790	855	915	39
34. Vertical grip reach (standing)	1980	2125	2270	87	1830	1970	2110	85
35. Vertical grip reach (sitting)	1145	1240	1330	57	1050	1145	1235	57
36. Forward grip reach	655	730	805	46	610	670	730	37

See notes on pp. 176–7.

Table 10.38 Anthropometric estimates for British 18-year-olds (all dimensions in mm).

	Men				Women			
Dimension	5th %ile	50th %ile	95th %ile	SD	5th %ile	50th %ile	95th %ile	SD
1. Stature	1660	1760	1860	60	1530	1620	1710	56*
2. Eye height	1555	1650	1745	59	1430	1520	1610	55
3. Shoulder height	1355	1445	1535	54	1235	1320	1405	52
4. Elbow height	1010	1105	1175	44	940	1010	1080	42
5. Hip height	865	935	1005	43	755	820	885	40
6. Knuckle height	705	765	825	35	670	725	780	32
7. Fingertip height	585	640	700	35	560	610	665	32
8. Sitting height	860	915	970	32	800	855	910	32
9. Sitting eye height	745	800	855	32	695	745	795	30
10. Sitting shoulder height	550	600	650	30	515	560	605	28
11. Sitting elbow height	200	245	290	26	185	230	275	26
12. Thigh thickness	135	160	185	15	120	145	170	14
13. Buttock–knee length	545	590	635	26	515	560	605	28
14. Buttock–popliteal length	450	500	550	29	435	480	525	27
15. Knee height	505	550	595	26	455	500	545	26
16. Popliteal height	405	445	485	25	365	405	445	25
17. Shoulder breadth (bideltoid)	415	455	495	23	360	395	430	21
18. Shoulder breadth (biacromial)	365	395	425	17	335	360	385	16
19. Hip breadth	300	340	380	25	300	345	390	27
20. Chest (bust) depth	190	225	260	21	195	235	275	24
21. Abdominal depth	205	240	275	21	185	220	255	20
22. Shoulder–elbow length	340	370	395	17	310	335	360	16
23. Elbow–fingertip length	450	480	510	18	395	425	455	17
24. Upper limb length	740	790	840	31	660	710	760	29
25. Shoulder–grip length	615	665	715	31	550	595	645	29
26. Head length	185	200	215	8	170	180	190	7
27. Head breadth	145	155	165	5	135	145	155	5
28. Hand length	175	190	205	8	160	175	190	8
29. Hand breadth	85	90	95	4	70	75	80	4
30. Foot length	250	270	290	12	220	240	260	11
31. Foot breadth	90	100	110	5	80	90	100	5
32. Span	1695	1810	1925	71	1520	1620	1720	62
33. Elbow span	890	955	1020	40	795	855	920	38
34. Vertical grip reach (standing)	2045	2150	2255	65	1830	1970	2110	85
35. Vertical grip reach (sitting)	1170	1250	1335	52	1065	1150	1235	52
36. Forward grip reach	675	740	805	41	610	670	730	37

See notes on pp. 176–7.

A mathematical synopsis of anthropometrics

A1   The normal distribution

Consider a variable x which is normally distributed in a population. Its probability density function is

$$f(x) = (1/\sigma\sqrt{2\pi}) \exp[-(x - \mu)^2/2\sigma^2] \tag{A1}$$

where μ is the mean and σ is the standard deviation of x for the population. $f(x)$ is a measure of the relative probability or relative frequency of the variable having a given value of x – it will be found in books of statistical tables as the 'ordinate of the normal curve'. If variable x is replaced by the standard normal deviate (z), such that

$$z = (x - \mu)/\sigma \tag{A2}$$

Equation A1 becomes

$$f(z) = (1/\sqrt{2\pi}) \exp(-z^2/2) \tag{A3}$$

which is known as the standardized form of the normal distribution (with zero mean and unit standard deviation).

The probability that x is less than or equal to a certain value is given by

$$F(x) = \int_{-\infty}^{x} f(x) \, \mathrm{d}x \tag{A4}$$

i.e. $F(x)$ corresponds to the area between the abscissa and the curve from $-\infty$ to x. This is the cumulative normal curve or normal ogive. It is tabulated in Table A1, in which $F(x)$ is given as a percentage (p) for given values of z (as defined above).

Table A1 p and z values of the normal distribution.

p	z	p	z	p	z	p	z
1	−2.33	26	−0.64	51	0.03	76	0.71
2	−2.05	27	−0.61	52	0.05	77	0.74
3	−1.88	28	−0.58	53	0.08	78	0.77
4	−1.75	29	−0.55	54	0.10	79	0.81
5	−1.64	30	−0.52	55	0.13	80	0.84
6	−1.55	31	−0.50	56	0.15	81	0.88
7	−1.48	32	−0.47	57	0.18	82	0.92
8	−1.41	33	−0.44	58	0.20	83	0.95
9	−1.34	34	−0.41	59	0.23	84	0.99
10	−1.28	35	−0.39	60	0.25	85	1.04
11	−1.23	36	−0.36	61	0.28	86	1.08
12	−1.18	37	−0.33	62	0.31	87	1.13
13	−1.13	38	−0.31	63	0.33	88	1.18
14	−1.08	39	−0.28	64	0.36	89	1.23
15	−1.04	40	−0.25	65	0.39	90	1.28
16	−0.99	41	−0.23	66	0.41	91	1.34
17	−0.95	42	−0.20	67	0.44	92	1.41
18	−0.92	43	−0.18	68	0.47	93	1.48
19	−0.88	44	−0.15	69	0.50	94	1.55
20	−0.84	45	−0.13	70	0.52	95	1.64
21	−0.81	46	−0.10	71	0.55	96	1.75
22	−0.77	47	−0.08	72	0.58	97	1.88
23	−0.74	48	−0.05	73	0.61	98	2.05
24	−0.71	49	−0.03	74	0.64	99	2.33
25	−0.67	50	0	75	0.67		

p	z	p	z
2.5	−1.96	97.5	1.96
0.5	−2.58	99.5	2.58
0.1	−3.09	99.9	3.09
0.01	−3.72	99.99	3.72
0.001	−4.26	99.999	4.26

A2   Samples, populations and errors

In reality we cannot know μ and σ, the parameters of a population (except in very special circumstances). We can only infer or estimate them from m and s – the mean and standard deviation of a sample of individuals deemed to be representative of the population, such that

$$m = \Sigma\, x/n \tag{A5}$$

and

$$s = \sqrt{\Sigma\, (x - m)^2/n} \tag{A6}$$

where n is the number of subjects in the sample.

In small samples (e.g. $n = 30$) it is conventional to make the arbitrary correction

$$s = \sqrt{\Sigma (x - m)^2/(n - 1)} \qquad (A7)$$

The quantity s is known as the variance of the sample.

In many cases it is more convenient to calculate the standard deviation by means of the identity

$$\Sigma(x - m)^2 = \Sigma x^2 - (\Sigma x)^2/n \qquad (A8)$$

As n increases, m and s become more reliable estimates of μ and σ, i.e. the likely magnitude of random sampling errors diminishes. (Note that we are not talking about errors of bias due to non-representative sampling – this is a more complex matter.) Sampling errors in estimating population parameters may be shown to be normally distributed with a mean of zero and a standard deviation known as the standard error (SE) of the parameter concerned, such that

$$SE\ mean = s/\sqrt{n} \qquad (A9)$$

$$SE\ standard\ deviation = s/\sqrt{2n} = 0.71\ SE\ mean \qquad (A10)$$

$$SE\ pth\ \%ile = \frac{p(100 - p)s}{100 f_p n} \qquad (A11)$$

where f_p is the ordinate of the normal curve at the pth %ile.

Probable magnitudes of sampling errors are commonly expressed in terms of the 95% confidence limits of the parameter concerned, which are calculated as ± 1.96 SE, i.e. the true values of a population parameter will lie within ± 1.96 standard errors of the sample statistic, 95 times out of every 100 that the sample is drawn. (Alternatively, if we are concerned with errors in one direction only, we use 1.645 SE.)

To simplify matters we may summarize this by saying that in any anthropometric survey the 95% confidence limits of a statistic ($\pm U_{95}$) are given by

$$U_{95} = ks/\sqrt{n} \qquad (A12)$$

where k is a constant for the statistic concerned given in Table A2, or, alternatively,

Table A2 Values of the parameter k, as given in equation A16	
Statistic	k
Mean	1.96
Standard deviation	1.39
Percentiles	
40th and 60th	2.49
30th and 70th	2.58
20th and 80th	2.80
10th and 90th	3.35
5th and 95th	4.14
1st and 99th	7.33

the equation

$$n = \left(\frac{ks}{U_{95}}\right)^2 \tag{A13}$$

gives us an indication of the number of subjects we need to measure in order for a particular statistic to give a certain desired degree of accuracy.

A3 The coefficient of variation

The coefficient of variation (CV) is given by

$$CV = s/m \times 100\% \tag{A14}$$

It is a useful index of the inherent variability of a dimension, i.e. it is independent both of absolute magnitude and of units of measurement.

In most populations stature has a lower CV than any other dimension. (Does this reflect a biological phenomenon or is it an artefact of measurement?) Characteristic ranges of CV of various types of anthropometric data are shown in Table A3. The figures were gathered from a number of sources (Damon *et al.* 1966, Roebuck *et al.* 1975, Grieve and Pheasant 1982) and do not reflect any specific population. They should rather be seen as a general guide to the approximate levels that we might anticipate. The high CVs of the lower part of the table are indicative of a skewed distribution – which is characteristic of anthropometric dimensions including soft tissue (fat) and of functional measures such as strength.

Roebuck *et al.* (1975) have demonstrated that for body length and breadth dimensions, in general, the relationship between standard deviation and mean will tend to be curvilinear (i.e. CV declines with increasing mean value). The reasons behind this observation are obscure but may be concerned with measurement error. Figure A1 shows this relationship plotted out for the 36 dimensions of Table 10.1.

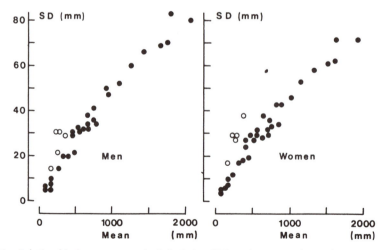

Figure A1   Relationship between standard deviation (SD) and mean in the anthropometric data of Table 10.1. ○ = body depths, thigh thickness, sitting elbow height and hip breadth; ● = all other dimensions.

Table A3 Characteristic coefficients of variation of anthropometric data.

Dimension	CV (%)
Stature	3–4
Body heights (sitting height, elbow height, etc.)	3–5
Parts of limbs	4–5
Body breadths (hips, shoulder, etc.)	5–9
Body depths (abdominal, chest, etc.)	6–9
Dynamic reach	4–11
Weight	10–21
Joint ranges	7–28
Muscular strength (static)	13–85

A4 Combining distributions

There are numerous situations in which the parameters of two or more normally distributed samples or populations must be combined to give a single lumped distribution. Strictly speaking, the new lumped distribution cannot be normal (Gaussian). To describe it we should calculate percentiles iteratively. Consider two sample distributions $m_1[s_1]$ and $m_2[s_2]$ of n_1 and n_2 subjects, respectively. For any value of x calculate standard normal deviates z_1 and z_2 in the two distributions and convert to percentiles p_1 and p_2 using Table A1. The percentile p' in the lumped distributions is given by

$$p' = \frac{p_1 n_1 + p_2 n_2}{n_1 + n_2} \tag{A15}$$

To describe the lumped distribution this process is repeated iteratively for as many values of x as is required.

However, in many cases the lumped distribution may be approximated by a new normal distribution $m'[s']$. In order to do this we must recalculate the sum (Σx) and the sum of squares (Σx^2) of the original raw data. For each normal distribution

$$\Sigma x = nm \tag{A16}$$

and from Equations A6 and A8

$$\Sigma x^2 = ns^2 + nm^2 \tag{A17}$$

and for k samples described by m_i, s_i, n_i, etc.

$$\Sigma x = \sum_{i=1}^{k} (n_i m_i) \tag{A18}$$

$$\Sigma x^2 = \sum_{i=1}^{k} (n_i s_i^2 + n_i m_i^2) \tag{A19}$$

and

$$n' = \sum_{i=1}^{k} (n_i) \tag{A20}$$

Therefore,

$$m' = \Sigma(n_i m_i)/\Sigma(n_i) \tag{A21}$$

$$(s')^2 = \frac{\{\Sigma(n_i s_i^2 + n_i m_i^2) - [\Sigma(n_i m_i)]^2/\Sigma n_i\}}{n_i} \tag{A22}$$

The validity of the latter approach is greatest when the constituent sample standard deviations are large and the differences between the means are small.

In the case when n is the same for all k samples then

$$m' = \Sigma m_i/k \tag{A23}$$

and

$$(s')^2 = \frac{\Sigma(m_i^2 + s_i^2) - (\Sigma m_i)^2/k}{k} \tag{A24}$$

A5 The bivariate distribution

Consider two normally distributed variables x and y. Their joint probability density function is given by

$$f_{(xy)} = 1/2\pi\sigma_x\sigma_y\sqrt{1 - \rho^2}\,e^w \tag{A25}$$

where

$$w = -(x - \mu_x)^2/\sigma_x - 2\rho(x - \mu_x)(y - \mu_y)/\sigma_x\sigma_y + \frac{(y - \mu_y)^2\sigma_y^2}{2(1 - \rho^2)} \tag{A26}$$

This distribution has five parameters of which μ_x, μ_y, μ_x, μ_y are self-explanatory and ρ is the correlation coefficient for the population which is best estimated by the sample correlation coefficient (r), where

$$r = \frac{\Sigma(x - m_x)(y - m_y)}{\sqrt{\Sigma(x - m_x)^2\Sigma(y - m_y)^2}} \tag{A27}$$

or

$$r = s_{xy}/s_x s_y \tag{A28}$$

where

$$s_{xy} = \frac{\Sigma(x - m_x)(y - m_y)}{n} \tag{A29}$$

which is known as the covariance of x and y and s_x and s_y are the sample standard deviations of x and y, respectively.

If the surface of the bivariate probability function is cut by a plane parallel to the y axis, the intersection will describe a normal probability curve. This curve defines the distribution of values of y as a population of subjects who all have the same value of x; and the mean of this distribution is the most probable value of y for a given value of x. The means of all such distributions fall on a straight line known as the regression line of y on x, given by the equation

$$y = a + bx \tag{A30}$$

where

$$b = \frac{\Sigma(x - m_x)(y - m_y)}{\Sigma(x - m_x)^2} = \frac{s_{xy}}{s_x^2} = \frac{rs_y}{s_x}$$ (A31)

and

$$a = m_y - bm_x$$ (A32)

(Similarly, the means of normal distributions, given by sections cutting the bivariate distribution parallel to the x axis, define the regression line of x and y. The two regression lines will be co-incident if r is 1; perpendicular if r is 0).

Equations A30–A32 may also be written

$$y - m_y = r(s_x/s_y)(x - m_x)$$ (A33)

At any given value of x, y is normally distributed with a mean defined by the regression line and a standard deviation given by

$$SE = s_y\sqrt{1 - r^2}$$ (A34)

(SE being known as the standard error of the estimate.)

A6 Estimating unknown distributions

The practical anthropometrist is frequently required to estimate the distribution of a dimension that, for reasons of practical expediency, may not be measured directly in a particular population. Some useful techniques will be described.

A6.1 Correlation and regression

If the parameters m_x, m_y, s_x, s_y and r are known in sample 1, and the parameters m_x and s_x are known in sample 2, then m_y and s_y may be estimated in sample 2 from the equations given above on the assumption that r is the same in both samples. (These samples may, of course, be deemed to be representative of populations.)

A6.2 Sum and difference dimensions

When an unknown dimension is anatomically equivalent to the sum of two known dimensions then

$$m_{(x+y)} = m_x + m_y$$ (A35)
$$s^2_{(x+y)} = s_x^2 + s_y^2 + 2rs_x s_y$$ (A36)

When an unknown dimension is anatomically equivalent to the difference between two known dimensions then

$$m_{(x-y)} = m_x - m_y$$ (A37)

$$s^2_{(x-y)} = s_x^2 + s_y^2 - 2rs_x s_y$$ (A38)

A6.3 The method of ratio scaling

If the parameters of variables x and y are known in a reference population a (or more precisely in a sample drawn from it) but only the parameters of x are known in population b (which we shall call the 'target population') then

$$m_y/m_x \text{ (in population a)} \simeq m_y/m_x \text{ (in population b)} \tag{A39}$$

and

$$s_y/s_x \text{ (in population a)} \simeq s_y/s_x \text{ (in population b)} \tag{A40}$$

provided that populations a and b are similar in age, sex, and ethnicity.

   Although these equations cannot be justified mathematically they have been widely employed, both in the present text and elsewhere, on grounds of practical expediency (e.g. Barkla 1961). We may call the dimension x, which is known in both populations, the scaling dimension – stature is most commonly used for this purpose since it is commonly available for populations in which other data are sparse. The simplest technique is to collect, from a variety of reference populations, the coefficients

$$E_1 = \frac{\text{mean of required dimension}}{\text{mean stature}} \tag{A41}$$

$$E_2 = \frac{\text{standard deviation at required dimension}}{\text{standard deviation of stature}} \tag{A42}$$

and simply multiply these by the parameters of stature in the target population. Pheasant (1982a) conducted a validation study of this technique and found that its errors were acceptable for most purposes.

A6.4 Empirical estimation of standard deviation

The data plotted in Figure A1 may be empirically fitted with the following regression equations.
   For body heights, lengths, and breadths:

$$\text{men: } s = 0.05703m - 0.000008347m^2 \tag{A43}$$

$$\text{women: } s = 0.05783m - 0.000010647m^2 \tag{A44}$$

For body depths, thigh thickness, sitting elbow height, and hip breadth:

$$\text{men: } s = 7.864 + 0.06977m \tag{A45}$$

$$\text{women: } s = 4.249 + 0.09467m \tag{A46}$$

   These equations may then be used as a first estimate of the standard deviation of a dimension for which the mean is known or can be calculated.
   Alternatively, if the coefficient of variation of a similar dimension is known, it may be assumed that the CV of the required dimension is the same and the standard deviation may again be calculated from the mean. (If Equations A43–A46 hold then this latter assumption will tend to overestimate the standard deviation of large dimensions and underestimate that of small ones.)

A7   Combinations of people

Consider distributions $m_a[s_a]$ and $m_b[s_b]$. (These might be the same variable in different samples of individuals or different variables in the same sample.)

If members of the two distributions meet at random (i.e. chance encounters occur) the distribution of differences is given by

$$m_{(a-b)} = m_a - m_b \tag{A47}$$

$$s^2_{(a-b)} = s^2_a + s^2_b \tag{A48}$$

and the distribution of sums by

$$m_{(a-b)} = m_a + m_b \tag{A49}$$

$$s^2_{(a+b)} = s^2_a + s^2_b \tag{A50}$$

In certain design applications it is necessary to know the breadth of two or more people placed side by side, e.g. upon a bench seat. If the body breadth concerned has the distribution $m[s]$ then the parameters of the group distribution are given by

$$m_g = nm \tag{A51}$$

$$s_g = s\sqrt{n} \tag{A52}$$

where n is the number of people in the group.

References and further reading

ABEYSEKERA, J. D. A. and SHANAVAZ, H. (1987) Body size of Sri Lankan workers and their variability with other populations in the world: its impact on the use of imported goods, *Journal of Human Ergology*, **16**, 193–208.

ABRAHAM, S. (1979) Weight and height of adults 18–74 years of age, United States, 1971–1974, *Vital and Health Statistics*, Series 11, No. 211, US Department of Health Education and Welfare, MD.

ADAMS, G. A. (1961) A comparative anthropometric study of hard labor during youth as a stimulator of physical growth of young colored women, *Research Quarterly AAHPER*, **9**, 102–8.

ANDERSSON, G. B. J. (1979) Low back pain in industry: epidemiological aspects, *Scandinavian Journal of Rehabilitation Medicine*, **II**, 163–8.

ANDERSSON, G. B. J., ORTENGREN, R., NACHEMSON, A. and ELFSTROM, G. (1974) Lumbar disc pressure and myoelectric back muscle activity during sitting. I, Studies on an experimental chair, *Scandinavian Journal of Rehabilitation Medicine*, **3**, 104–14.

ANDERSSON, M., HWANG, S. G. and GREEN, W. T. (1965) Growth of the normal trunk in boys and girls during the second decade of life, *Journal of Bone and Joint Surgery*, **47A**, 1554–64.

ANDREW, I. and MANOY, R. (1972) Anthropometric survey of British Rail footplate staff, *Applied Ergonomics*, **3**, 132–5.

ARCHER, J. and LLOYD, B. (1982) *Sex and Gender*, Harmondsworth: Penguin.

ARMSTRONG, T. J., BUCKLE, P., FINE, L. J., HAGBERG, M., JONSSON, B., KILBLOM, A., KUORINKA, I. A. A., SILVERSTEIN, B. A., SJOGAARD, G. and VIIKARI-JUNTURA, E. R. A. (1993) A conceptual model for work-related neck and upper-limb musculoskeletal disorders, *Scandinavian Journal of Work Environment and Health*, **19**, 73–84.

ASHLEY-MONTAGU, M. F. (1960) *An Introduction to Physical Anthropology*, 3rd Edn, Springfield, IL: Charles C. Thomas.

ASMUSSEN, E. and HEEBOLL-NIELSEN, K. (1962) Isometric muscle strength in relation to age in men and women, *Ergonomics*, **5**, 167–76.

ATHERTON, J., CLARKE, A. K. and HARRISON, E. '4. Office Seating for the Arthritic and Low Back Pain Patients.' Royal National Hospital for Rheumatic Diseases, Bath.

AYOUB, M. M. and McDANIEL, J. W. (1973) Effects of operator stance on pushing and pulling tasks, *AIIE Transactions*, **6**, 185–95.

BABBS, F. W. (1979) A design layout method for relating seating to the occupant and vehicle, *Ergonomics*, **22**, 227–34.

BACKWIN, H. and MCLAUGHLIN, S. D. (1964) Increase in stature – is the end in sight? *Lancet*, **ii**, 1195–7.

BALLANTINE, M. (1983) Well, how do children learn population stereotypes? *Proceedings of the Ergonomics Society's Conference 1983*, ed. K. COOMBES, p. 1, London: Taylor & Francis.

BARKLA, D. (1961) The estimation of body measurements of the British population in relation to seat design, *Ergonomics*, **4**, 123–32.

BARNES, R. M. (1958) *Motion and Time Study*, New York: Wiley.

BARTER, T., EMMANUEL, I. and TRUETT, B. (1957) A statistical evaluation of joint range data, WADC Technical Note 53-311, Wright Patterson Airforce Base, OH.

BATOGOWSKA, A. and SLOWIKOWSKI, J. (1974) *Anthropometric Atlas of the Polish Adult Population for Designer Use*, Warsaw: Instytut Wzornictwa Przemystowego (in Polish).

BAUER, D. and CAVONIUS, C. R. (1980) Improving the legibility of visual display units through contrast reversal, in Grandjean and Vigliani (1980), pp. 137–42.

BENDIX, T. and BIERING-SORENSEN, F. (1983) Posture of the trunk when sitting on forward inclining seats, *Scandinavian Journal of Rehabilitation Medicine*, **15**, 197–203.

BENDIX, T. and HAGBERG, M. (1984) Trunk posture and load on the trapezius muscle whilst sitting at sloping desks, *Ergonomics*, **27**, 873–82.

BENNETT, C. (1977) *Spaces for People: Human Factors in Design*, Englewood Cliffs, NJ: Prentice Hall.

BITTNER A. C., DANNHAUS, D. M. and ROTH, J. T. (1975) Workplace-accommodated percentage evaluation: model and preliminary results, in M. M. AYOUB and C. G. HALCOMB (Eds) *Improved Seat, Console and Workplace Design*, Pacific Missile Test Center, Point Mugu, CA 93042.

BOAS, F. (1912) *Changes in Bodily Form of Descendants of Immigrants*, New York: Columbia University Press.

BORKAN, G. A., HULTS, D. E. and GLYNN, R. J. (1983) Role of longitudinal change and secular trend in age differences in male body dimensions, *Human Biology*, **55**, 629–41.

BORKAN, G. A. and NORRIS, A. H. (1977) Fat redistribution and the changing body dimensions of the adult male, *Human Biology*, **49**, 495–514.

BRANTON, P. (1984) Backshapes of seated persons: how close can the interface be designed?, *Applied Ergonomics*, **15**, 105–7.

BROWN, C. H. and WILMORE, J. H. (1974) The effects of maximal resistance training on the strength and body composition of women athletes, *Medicine and Science in Sports*, **6**, 174–7.

BROWN, C. R. and SCHAUM, D. L. (1980) User-adjusted VDU parameters, in Grandjean and Vigliani (1980), pp. 195–100.

BRUNSWIC, M. (1981) 'How seat design and task affect the posture of the spine in unsupported sitting', MSc. dissertation, Ergonomics Unit, University College London.

BS 3042: 1971 (1980) *Standard Test Fingers and Probes for Checking Protection against Electrical, Mechanical and Thermal Hazards*, London: British Standards Institution.

BS 3693: Part 1 (1964) *Instruments for Bold Presentation and Rapid Reading*, London: British Standards Institution.

BS 3705 (1972) *Recommendations for Provision of Space for Domestic Kitchen Equipment*, London: British Standards Institution.

BS 5304 (1975) *Code of Practice for Safeguarding of Machinery*, London: British Standards Institution.

BS 5555 (1981) *Specification for SI Units and Recommendations for the Use of their Multiples and Certain Other Units (= ISO 1000)*, London: British Standards Institution.

BS 5619 (1978) *Code of Practice for Design of Housing for the Convenience of Disabled People*, London: British Standards Institution.

BS 5775: Part O (1982) *Specification for Quantities, Units and Symbols* (= *ISO 31/0*), London: British Standards Institution.

BS 5940: Part I (1980) *Office Furniture: Specification for Design and Dimensions of Office Workstations, Desks, Tables and Chairs*, London: British Standards Institution.

BS 5959 (1980) *Specification for Key Numbering System and Layout Charts for Keyboards on Office Machines* (= *ISO 4169*), London: British Standards Institution.

BS 6222: Part I (1982) *Domestic Kitchen Equipment: Specification for Co-ordinating Dimensions*, London: British Standards Institution.

BS AU176 (1980) *Method for Establishment of Eyellipses for Drivers's Eye Location* (= *ISO 4513*), London: British Standards Institution.

CAILLET, R. (1977) *Soft Tissue Pain and Disability*, Philadelphia: F. A. Davis.

CAKIR, A., HART, D. J. and STEWART, T. F. M. (1980) *Visual Display Terminals*, Chichester: Wiley.

CAMERON, I. (1984) 'What is common sense?', unpublished paper presented to the London Ergonomics Group.

CAMERON, N. (1979) The growth of London schoolchildren 1904–1966: an analysis of secular trend and intra-county variation, *Annals of Human Biology*, **6**, 505–25.

CAMERON, N., TANNER, J. M. and WHITEHOUSE, R. A. (1982) A longitudinal analysis of the growth of limb segments in adolescence, *Annals of Human Biology*, **9**, 211–20.

CANNON, L. J., BERNACKI, E. J. and WALTER, S. D. (1981) Personal and occupational risk factors associated with the carpal tunnel syndrome, *Journal of Occupational Medicine*, 255–8.

CHAFFIN, D. B. and ANDERSSON, G. B. J. (1984) *Occupational Biomechanics*, New York: Wiley.

CHAPANIS, A. (1959) *Research Techniques in Human Engineering*, Baltimore: The Johns Hopkins Press.

CHAPANIS, A. and LINDENBAUM, L. (1950) A reaction time study of four central display linkages, *Human Factors*, **I**, 1–17.

CHINN, S., RONA, R. J., PRICE, C. E. (1989) The secular trend in height of primary school children in England and Scotland, 1972–1979 and 1979–1986, *Annals of Human Biology*, **16**, 387–395.

CLARK, T. S. and CORLETT, E. N. (1984) *The Ergonomics of Workspaces and Machines: A Design Manual*, London: Taylor & Francis.

COHEN, M. L., ARROYO, J. F., CHAMPION, G. D. and BROWNE, C. D. (1992) In search of the pathogenesis of refractory cervicobrachial pain syndrome: a deconstruction of the RSI phenomenon, *The Medical Journal of Australia*, **156**, 432–6.

CONRAD, R. and HULL, A. J (1968) The preferred layout for numeral data entry keysets, *Ergonomics*, **II**, 165–74.

COOPER, C., MCALINDON, T., COGGON, D., EGGER, P. and DIEPPE, P. (1994) Occupational activity and osteoarthritis of the knee, *Annals of the Rheumatic Diseases*, **53**, 90–3.

CORLETT, E. N. (1983) Analysis and evaluation of working posture, in T. O. KVALSETH (Ed.) *Ergonomics of Workstation Design*, London: Butterworths.

CORLETT, E. N. and MANENICA, I. (1980) The effects and measurement of working postures, *Applied Ergonomics*, **II**, 7–16.

COX, C. F. (1984) 'An investigation of the dynamic anthropometry of the seated workplace', MSc. dissertation, University College London.

CYRIAX, J. H. (1978) *Textbook of Orthopaedic Medicine*, 7th Edn, London: Baillière Tindall.

DALASSIO, D. J. (Ed.) (1980) *Wolff's Headache and Other Head Pain*, Oxford: Oxford University Press.

DAMON, A. (1973) Ongoing human evolution, in I. H. PORTER and R. E. SKALKO (Eds), *Heredity and Society*, pp. 45–74, New York: Academic Press.

DAMON, A., STOUDT, H. W. and MCFARLAND, R. A. (1966) *The Human Body in*

Equipment Design, Cambridge MA: Harvard University Press.

DAMON, A., SELTZER, C. C., STOUDT, H. W. and BELL, B. (1972) Age and physique in healthy white veterans at Boston, *Journal of Gerontology*, **27**, 202–8.

DAS, B. and GRADY, R. M. (1983) The normal working area in the horizontal plane: a comparative analysis between Farley's and Squires' concepts, *Ergonomics*, **26**, 449–59.

DAVID, G. C. (1987) Intra-abdominal pressure measurements and load capacities for females, *Ergonomics*, **28**, 345–8.

DAVIES, B. T., ABADA, A., BENSON, K., COURTNEY A. and MINTO, I. (1980a) Female hand dimensions and guarding of machines, *Ergonomics*, **23**, 79–84.

DAVIES, B. T., ABADA, A., BENSON, K., COURTNEY, A. and MINTO, I. (1980b) A comparison of hand anthropometry of females in three ethnic groups, *Ergonomics*, **23**, 179–82.

DAVIS, P. R. and STUBBS, D. A. (1977) Safe levels of manual forces for young males, *Applied Ergonomics*, **8**, 141–50, 219–28.

DEMPSTER, W. T. (1955) Space requirements of the seated operator: geometrical, kinematic and mechanical aspects of the body with special reference to the limbs, WADC Tech. Note 55-159, Wright Patterson Air Force Base, OH.

DEPARTMENT OF DEFENSE (1981) *Human Engineering Design Criteria for Military Systems, Equipment and Facilities*, MIL-STD-1472C, Washington, DC: Department of Defense.

DEPARTMENT OF EDUCATION AND SCIENCE (1972) *British School Population Dimensional Survey*, Building Bulletin 46, Department of Education and Science, London: HMSO.

DEPARTMENT OF EDUCATION AND SCIENCE (1985) *Body Dimensions of the School Population*, Building Bulletin 62, Department of Education and Science, London: HMSO.

DEPARTMENT OF THE ENVIRONMENT (1972) *Space in the Home*, Design Bulletin 6, Department of the Environment, London: HMSO.

DIFFRIENT, N., TILLEY, A. and BARDAGIY, J. C. (1974) *Humanscale 1/2/3*, Cambridge, MA: MIT Press.

DIFFRIENT N., TILLEY, A. and HARMAN, D. (1981) *Humanscale 4/5/6/7/8/9*, Cambridge, MA: MIT Press.

DIN 31001 Part I (1983) *Safety Distances for Adults and Children*, Berlin: Deutsches Institut für Normung.

DREYFUSS, H. (1971) *The Measure of Man*, New York: Whitney.

DRILLIS, R. and CONTINI, R. (1966) Body segment parameters, Technical Report No. 1166.03, School of Engineering Science, New York University.

DRURY, C. G. (1980) Handles for manual materials handling, *Applied Ergonomics*, **II**, 35–42.

DRURY, C. G. and FRANCHER, M. (1985) Evaluation of a forward-sloping chair, *Applied Ergonomics*, **16**, 41–7.

DUCHARME, R. E. (1975) Problem tools for women, *Industrial Engineering*, **7**, 46–50.

DUCHARME, R. E. (1977) Women workers rate male tools inadequate, *Human Factors Society Bulletin*, **20**, 1–2.

DUNCAN, J. and FERGUSON, D. (1974) Keyboard operating posture and symptoms in operating, *Ergonomics*, **17**, 651–62.

DURNIN, J. V. G. A. and RAHAMAN, M. M. (1967) The assessment of the amount of fat in the human body from measurements of skinfold thickness, *British Journal of Nutrition*, **21**, 681–9.

DURNIN, J. V. G. A. and WOMERSLEY, J. (1974) Body fat assessed from total body density and its estimation from skinfold thickness: measurements on 481 men and women aged from 16 to 72 years, *British Journal of Nutrition*, **32**, 77–97.

ELMFELDT, G., WISE, C., BERESTEN, H. and OLSSON, A. (1981) *Adapting Work Sites for People with Disabilities*, The Swedish Institute for the Handicapped, S-16126 Bromma, Sweden/World Rehabilitation Fund, 400 East 34th Street, New York.

ERICSSON (1983) *Ergonomic Principles in Office Automation*, Bromma: Ericsson Information Systems AB.

EVELETH, P. B. (1975) Differences between ethnic groups in sex dimorphism of adult height, *Annals of Human Biology*, **2**, 35–9.

EVELETH, P. B. and TANNER, J. M. (1976) *Worldwide Variation in Human Growth*, Cambridge: Cambridge University Press.

FLOYD, W. F., GUTTMAN, L., WYCLIFFE-NOBLE, C., PARKES, M. A. and WARD, B. A. (1966) A study of the space requirements of wheelchair users, *Paraplegia*, May 34–7.

FLOYD, W. F. and WARD, J. S. (1966) Posture in industry, *International Journal of Production Research*, **5**, 213–24.

FRIEDLANDER, J. S., COSTA, P. T., BOSSE, R., ELLIS, E., RHOADS, J. G. and STOUDT, H. W. (1977) Longitudinal physique changes among healthy white veterans at Boston, *Human Biology*, **49**, 541–58.

FRUIN, J. (1971) *Pedestrian Planning and Design*, New York: Metropolitan Association of Urban Designers and Environmental Planners.

GALER, D. M., BAINES, A. and SIMMONDS, G. (1980) Ergonomic aspects of electronic dashboard instrumentation, in Oborne and Levis (1980), Vol. I, pp. 301–9.

GARNER, D. M., GARFINKLE, P. E., SCHWARTZ, D. and THOMPSON, M. (1980) Cultural expectations of thinness in women, *Psychological Reports*, **47**, 483–91.

GARRET, J. W. (1971) The adult human hand: some anthropometric and biomechanical considerations, *Human Factors*, **13**, 117–31.

GIBSON, S. J., LE VASSEUR, S. A. and HELME, R. D. (1991) Cerebral event-related responses induced by CO_2 laser stimulation in subjects suffering from cervico-brachial syndrome, *Pain*, **47**, 173–82.

GITE, L. P. and YADAV, B. G. (1989) Anthropometric survey for agricultural machine design, *Applied Ergonomics*, **20**, 191–6.

GOLDSMITH, S. (1976) *Designing for the Disabled*, 3rd Edn, London: RIBA.

GOLDSTEIN, H. (1971) Factors influencing the height of 7 year old children, *Human Biology*, **43**, 92-III.

GOODERSON, C. Y. and BEEBEE, M. (1976) Anthropometry of 500 infantrymen 1973–1974, Report APRE 17/76, Army Personnel Research Establishment, Farnborough, Hants.

GOODERSON, C. Y. and BEEBEE, M. (1977) A comparison of the anthropometry of 100 guardsmen with that of 500 infantrymen, 500 RAC servicemen and 2000 RAF aircrew, Report APRE 37/76, Army Personnel Research Establishment, Farnborough, Hants.

GOODERSON, C. Y., KNOWLES, D. J. and GOODERSON, P. M. E. (1982) The hand anthropometry of male and female military personnel, APRE Memorandum 82M510, Army Personnel Research Establishment, Farnborough, Hants.

GOULD, S. J. (1984) *The Mismeasure of Man*, Harmondsworth: Penguin.

GRAHAME, R. and JENKINS, J. M. (1972) Joint hypermobility: asset or liability? A study of joint mobility in ballet dancers, *Annals of the Rheumatic Diseases*, **31**, 109–11.

GRANDJEAN, E. (1973) *Ergonomics of the Home*, London: Taylor & Francis.

GRANDJEAN, E. (1981) *Fitting the Task to the Man: An Ergonomic Approach*, 2nd Edn, London: Taylor & Francis.

GRANDJEAN, E. (1986) *Fitting the Task to the Man: A Textbook of Occupational Ergonomics*, 4th Edn, London: Taylor & Francis.

GRANDJEAN, E. (1987) *The Ergonomics of Computerized Offices*, London: Taylor & Francis.

GRANDJEAN, E. and HÜNTING, W. (1977) Ergonomics of posture: review of various problems of standing and sitting posture, *Applied Ergonomics*, **8**, 135–40.

GRANDJEAN, E., HÜNTING, W., MAEDA, K. and LÄUBLI, TH. (1983) Constrained postures at office workstations, in T. O. KVALSETH (Ed.) *Ergonomics of Workstation Design*, London: Butterworths.

GRANDJEAN, E., HÜNTING, W. and NISHIYAMA, K. (1984) Preferred VDT workstation settings, body posture and physical impairments, *Applied Ergonomics*, **15**, 99–104.

GRANDJEAN, E., NISHIYAMA, K., HÜNTING, W. and PIDERMAN, M. (1984) A laboratory study on preferred and imposed settings of a VDT workstation, *Behaviour and Information Technology*, **3**, 289–304.

GRANDJEAN, E. and VIGLIANI, E. (Eds) (1980) *Ergonomic Aspects of Visual Display Terminals*, London: Taylor & Francis.

GREEN, R. A. and BRIGGS, C. A. (1989) Anthropometric dimensions and overuse injury among Australian keyboard operators, *Journal of Occupations Medicine*, **31**, 747–50.

GREULICH, H. (1957) A comparison of the physical growth and development of American born and native Japanese children, *American Journal of Physical Anthropology*, **15**, 489–516.

GRIEVE, D. W. (1984) The influence of posture on power output generated in single pulling movements, *Applied Ergonomics*, **15**, 115–16.

GRIEVE, D. W. and PHEASANT, S. T. (1982) Biomechanics, in W. T. SINGLETON (Ed.) *The Body at Work*, Cambridge: Cambridge University Press.

GRIMSHAW V. FORD MOTOR CO. (1981) 119 Cal App 3d 757. Cited in JONES, M. A. (1986) *Textbook on Torts*, London: Blackstone.

HAIGH, R. (1984) An ergonomic assessment of British Standard 5810: 1979: access for the disabled to buildings, through a survey of architects, Project Report, Department of Human Sciences, Loughborough University of Technology.

HALL, E. T. (1969) *The Hidden Dimension*, New York: Doubleday.

HAMER, M. (1985) How speed kills on our roads, *New Scientist*, 21 February, 10–11.

HAMILTON, N. (1983) 'Optimal location of ovens for wheelchair users: an ergonomic approach', MSc. dissertation, Ergonomics Unit, University College London.

HAMMOND, D. C. and ROE, R. W. (1972) SAE controls reach study, 73061, Society of Automotive Engineers.

HANSEN, R. and CORNOG, D. Y. (1958) Annotated bibliography of applied physical anthropology in human engineering, WADC Technical Report 56-30, Wright Patterson Airforce Base, OH.

HARRIS, A. I. (1971) *Handicapped and Impaired in Great Britain: Part I*, London: HMSO.

HARTLEY, J. and BURNHILL, P. (1977) Fifty guidelines for improving instructional text, *Programmed Learning and Educational Technology*, **14**, 65–73.

HASLEGRAVE, C. M. (1979) An anthropometric survey of British drivers, *Ergonomics*, **22**, 145–54.

HASLEGRAVE, C. M. and HARDY, R. N. (1979) *Anthropometric Profile of the British Car Occupant*, Motor Industry Research Association.

HASLEGRAVE, C. M. and SEARLE, J. A. (1980) Anthropometric considerations in seat belt design and testing, in Oborne and Levis (1980), Vol. I, pp. 374–82.

HEALTH AND SAFETY COMMISSION (1982) *Proposals for Health and Safety (Manual Handling of Loads): Regulations and Guidance*, Health and Safety Commission Consultative Document, London: HMSO.

HEALTH AND SAFETY EXECUTIVE (1975–1980a) *Health and Safety: Manufacturing and Service Industries*, London: HMSO.

HEALTH AND SAFETY EXECUTIVE (1978–1980b) *Health and Safety Statistics*, London: HMSO.

HEALTH AND SAFETY EXECUTIVE (1983) *Visual Display Units*, London: HMSO.

HEALTH AND SAFETY EXECUTIVE (1992a) *Display Screen Equipment Work: Guidance on Regulations*, London: HMSO.

HEALTH AND SAFETY EXECUTIVE (1992b) *Manual Handling: Guidance on Regulations*, London: HMSO.

HEALY, M. J. R. (1962) The effect of age-grouping on the distribution of a measurement effected by growth, *American Journal of Physical Anthropology*, **20**, 49–50.

HELME, R. D., LeVASSEUR, S. A. and GIBSON, S. J. (1992) RSI revisited: evidence for psychological and physiological differences from an age, sex and occupation matched control group, *Australian and New Zealand Journal of Medicine*, **22**, 23–9.

HERTZBERG, H. T. E. (1968) The conference on standardization of anthropometric techniques and terminology, *American Journal of Physical Anthropology*, **28**, 1–16.

HERTZBERG, H. T. E., CHURCHILL, E., DUPERTUIS, C. W., WHITE, R. M. and DAMON, A. (1963) *Anthropometric Survey of Greece, Turkey and Italy*, Oxford: Pergamon.

HETTINGER, T. (1961) *Physiology of Strength*, Springfield, IL: Charles C. Thomas.

HFDNE (1971) *Human Factors for Designers of Naval Equipment*, London: Admiralty.

HIRA, D. S. (1980) An ergonomic appraisal of educational desks, *Ergonomics*, **23**, 213–21.

HM CHIEF INSPECTOR OF FACTORIES (1924–1974) *Annual Report*, London: HMSO.

HOGARTH, W. (1753) *The Analysis of Beauty, Written with a View to Fixing the Fluctuating Ideas of Taste*, London.

HOPKINS, A. (1990) Stress, the quality of work and repetition strain injury in Australia, *Work and Stress*, **4**, 129–38.

HORNIBROOK, F. A. (1934) *The Culture of the Abdomen*, New York: Doubleday.

HSE (1982) *The Lifting of Patients in the Health Service*, London: HMSO.

HSE (1991) *Seating at Work*, HS (G) 60, London: HMSO.

HSE (1992) *Manual Handling: Guidance on Regulations*, London: HMSO.

HULT, L. (1954) Cervical, dorsal and lumbar spinal syndromes, *Acta Orthopaedica Scandinavia*, Supplement 17.

HUNTER, D. (1955) *Diseases of the Occupations*, London: English Universities Press.

HÜNTING, W., GRANDJEAN, E. and MAEDA, K. (1980) Constrained postures in accounting machine operators, *Applied Ergonomics*, **14**, 145–9.

HÜNTING, W., LÄUBLI, TH. and GRANDJEAN, E. (1981) Postural and visual loads at VDT workplaces. I, Constrained postures, *Ergonomics*, **24**, 917–31.

IKAI, M. and FUKANAGA, T. (1968) Calculation of muscle strength per unit cross-sectioned area of human muscle by means of ultrasonic measurement, *Int. Z. Angew, Physiol. Einschl. Arbeits-Physiol.*, **26**, 26–32.

ILO (1990) *Maximum Weights in Load Lifting and Carrying*, Occupational Safety and Health Series No. 59, International Labour Office, Geneva.

INGLEMARK, B. E. and LEWIN, T. (1968) Anthropometrical studies on Swedish women, *Acta Morphologica Neerlando-Scandinavica*, **III** (2), 145–66.

INSTITUTE FOR CONSUMER ERGONOMICS (1983) Seating for elderly and disabled people, Report No. 2, Anthropometric Survey, University of Technology, Loughborough.

INSTITUTO NACIONAL DE TECNOLOGIA (1989) *Pesquisa Antropometrica e Biomecacica dos Operarios da Industria Transformmacao – RJ*, Rio de Janeiro: Instituto Nacional De Tecnologia.

ISO 31/0 (1981) *General Principles Concerning Quantities, Units and Symbols* (= BSS 5775), Geneva: International Standards Organization.

ISO 3958 (1977) *Road Vehicles – Passenger Cars – Driver Hand Control Reach*, Geneva: International Standards Organization.

ISO 4040 (1983) *Road Vehicles – Passenger Cars – Location of Hand Controls, Indicators and Tell-tales*, Geneva: International Standards Organization.

ISO 4169 (1979) *Office Machines – Keyboards – Key Numbering System and Layout Charts* (= BS 5959), Geneva: International Standards Organization.

ISO 4513 (1978) *Road Vehicles – Visibility – Method for Establishment of Eyellipses for Drivers' Eye Location* (= BS AU176), Geneva: International Standards Organization.

ISO 6549 (1980) *Road Vehicles – Procedure for H-point Determination*, Geneva: International Standards Organization.

JAMES, J. (1951) A preliminary study of the size determinant in small group interaction, *American Sociological Review*, **16**, 474–7.

JONES, J. C. (1963) Fitting trials: a method of fitting equipment dimensions to variation in the activities, comfort requirements and body sizes of users, *The Architects Journal*, 6 February, 321–5.

JONES, J. C. (1969) Methods and results of seating research, *Ergonomics*, **12**, 171–81.

KAPANJI, I. A. (1974) *The Physiology of the Joints*, Edinburgh: Churchill Livingstone.

KAPLAN, B. A. (1954) Environment and human plasticity, *American Anthropology*, **56**, 780–800.

KARAGELIS, I. (1982) 'An ergonomic evaluation of proposed kitchen worktops for wheel-chair disabled', MSc. dissertation, Ergonomics Unit, University College London.

KEMBER, P., AINSWORTH, L. and BRIGHTMAN, P. (1981) *A Hand Anthropometric Survey of British Workers*, Ergonomics Laboratory, Cranfield Institute of Technology.

KENNEDY, K. W. (1964) Reach capability of the USAF population: Phase I, the outer boundaries of grasping reach envelopes for the shirt sleeved, seated operator, Report AMRL-TDR-64-59, Wright Patterson Airforce Base, OH.

KIRA, A. (1976) *The Bathroom*, Harmondsworth: Penguin.

KLAFS, C. E. and LYON, M. J. (1978) *The Female Athlete: A Coach's Guide to Conditioning and Training*, St Louis: C. V. Mosby.

KNIGHT, I. (1984) *The Heights and Weights of Adults in Great Britain*, London: HMSO.

KOBLIANSKI, E. and ARENSBURG, B. (1977) Changes in morphology of human populations due to migration and selection, *Annals of Human Biology*, **4**, 57–71.

KROEMER, K. H. E. (1971) Foot operation of controls, *Ergonomics*, **14**, 333–61.

KROEMER, K. H. E. (1972) Human engineering: the keyboard, *Human Factors*, **14**, 51–63.

KUKKONEN, R., LUOPAJARVI, T. and RIIHIMAKI, V. (1983) Prevention of fatigue amongst data entry operators, in T. O. KVALSETH (Ed.), *Ergonomics in Workstation Design*, London: Butterworths.

LAWSON, B. (1980) *How Designers Think*, London: Architectural Press.

LE CARPENTIER, E. F. (1969) Easy chair dimensions for comfort, *Ergonomics*, **12**, 328–37.

LE CORBUSIER (1961) *The Modulor: A Harmonious Measure to the Human Scale, Universally Applicable to Architecture and Mechanics*, London: Faber & Faber.

LEHMAN, G. (1958) Physiological basis of tractor design, *Ergonomics*, **I**, 197–205.

LEWIN, T. (1969) Anthropometric studies on Swedish industrial workers when standing and sitting, *Ergonomics*, **12**, 883–902.

LIFE, M. A. and PHEASANT, S. T. (1984) An integrated approach to the study of posture in keyboard operation, *Applied Ergonomics*, **15**, 83–90.

LINDGREN, G. (1976) Height, weight and menarche in Swedish schoolchildren in relation to socio-economic and regional factors, *Annals of Human Biology*, **3**, 510–28.

LITTLE, K. B. (1965) Personal space, *Journal of Experimental Social Psychology*, **I**, 237–47.

LOVELESS, N. E. (1962) Directional motion stereotypes: a review, *Ergonomics*, **5**, 357–81.

LUNDERVOLD, A. (1958) Electromyographic investigations during typewriting, *Ergonomics*, **1**, 226–33.

MCCLELLAND, I. L. and WARD, J. S. (1976) Ergonomics in relation to sanitary ware design, *Ergonomics*, **19**, 465–78.

MCCLELLAND, I. L. and WARD, J. S. (1972) The ergonomics of toilet seats, *Human Factors*, **24**, 713–25.

MCCORMICK, E. J. (1970) *Human Factors Engineering*, New York: McGraw-Hill.

MCWHIRTER, N. D. (1984) *Guinness Book of Records*, 30th Edn, Enfield: Guinness Super-latives.

MAEDA, K. (1977) Occupational cervicobrachial disorder and its causative factors, *Journal of Human Ergology*, **6**, 193–202.

MAGORA, A. (1972) Investigations of the relation between low back pain and occupation III. Physical requirements: sitting, standing and weight lifting, *Industrial Medicine*, **41**, 5–9.

MAGORA, A. (1973a) Investigations of the relation between low back pain and occupation IV. Physical requirements: bending, rotation, reaching and sudden maximal effort, *Scandinavian Journal of Rehabilitation Medicine*, **5**, 186–90.

MAGORA, A. (1973b) Investigations of the relation between low back pain and occupation

V. Psychological aspects, *Scandinavian Journal of Rehabilitation Medicine*, **5**, 191–6.

MALINA, R. M. and ZAVALETA, A. N. (1976) Androgyny of physique in female track and field athletes, *Annals of Human Biology*, **3**, 441–6.

MANDAL, A. C. (1976) Work chair with tilted seat, *Ergonomics*, **19**, 157–64.

MANDAL, A. C. (1981) The seated man (Homo sedens): the seated work position, theory and practice, *Applied Ergonomics*, **12**, 19–26.

MARQUER, P. and CHALMA, M. C. (1961) L'évolution des charactères morphologiques en function de l'âge chez 2089 françaises de 20 à 91 ans, *Bulletin et Mémoires de la Société d'Anthropologie de Paris*, **XI** (2), 1–78.

MARTIN, W. E. (1960) Children's body measurements for planning and equipping schools, Special Publication No. 4, US Department of Health, Education and Welfare, MD.

MARTIN, J. and WEBB, R. D. G. (1980) Response stereotypes for key turning, *Proceedings of the Annual Conference of the Human Factors Association of Canada*, 105–7.

MARTIN, J., MELTZER, H., ELIOT, D. (1988) *The Prevalence of Disability Among Adults*, London: HMSO.

MEDAWAR, P. B. (1944) Size, shape and age, in *Essays on Growth and Form Presented to D'Arcy Wentworth Thompson*, pp. 155–87, Oxford: Clarendon Press.

MELZACK, R. and WALL, P. (1982) *The Challenge of Pain*, Harmondsworth: Penguin.

MEREDITH, H. W. (1976) Findings from Asia, Australia, Europe and North America on secular change in mean height of children, youths and young adults, *American Journal of Physical Anthropology*, **44**, 315–26.

MIALL, W. E., ASHCROFT, M. T., LOVELL, H. G. and MOORE, F. (1967) A longitudinal study of the decline of adult height with age in two Welsh communities, *Human Biology*, **39**, 445–54.

MICHAELIS, P. R. (1980) An ergonomist's introduction to synthesized speech, in Oborne and Levis (1980), Vol. I, pp. 291–4.

MIIS (1984) Motor industry information service fact sheets, in *The Automania Fact File*, Birmingham: Central Television.

MILLER, C. D. (1961) Stature and build of Hawaii-born youth of Japanese ancestry, *American Journal of Physical Anthropology*, **19**, 159–71.

MILLER, D. I. and NELSON, R. C. (1973) *Biomechanics of Sport*, Philadelphia: Lea & Febiger.

MINISTRY OF DEFENCE (1984) *Defence Standard. Human Factors for Designers of Equipment, Part 3: Body Strength and Stamina*, MoD 00-25, Part 3, London: Ministry of Defence.

MITCHIE, D. and JOHNSTON, R. (1984) *The Creative Computer: Machine Intelligence and Human Knowledge*, Harmondsworth: Penguin.

MONTEGRIFFO, V. M. E. (1968) Height and weight of a United Kingdom adult population with a review of anthropometric literature, *Annals of Human Genetics*, **31**, 389–98.

MONTOYE, H. J. and LAMPHIER, D. E. (1977) Grip and arm strengths in males and females, *Research Quarterly of the American Association for Health, Physical Education and Recreation*, **48**, 109–20.

MORRIS, J. N., HEADY, J. A. and RAFFLE, P. A. B. (1956) Physique of London busmen: epidemiology of uniforms, *Lancet*, 15 September, 569–70.

MURREL, K. F. H. (1969) *Ergonomics: Man and His Working Environment*, London: Chapman & Hall.

NAPIER, J. R. (1956) The prehensile movements of the human hand, *Journal of Bone and Joint Surgery*, **38B**, 902–13.

NASA (1978) *Anthropometric Source Book*, NASA Defence Publication No. 1024, US National Aeronautics and Space Administration.

NICHOLSON, A. S., PARNELL, J. W. and DAVIS, P. R. (1985) Bed design for back pain sufferers, in *Ergonomics International 85*.

NIOSH (1981) *Work Practices Guide for Manual Lifting*, Cincinatti, OH: National Institute for Occupational Safety and Health.

NOBLE, J. (1982) *Activity and Spaces: Dimensional Data for Housing Design*, London: The Architectural Press.

NORFOLK, D. (1993) Bedding design: a professional appraisal, *British Osteopathic Journal*, **12**, 13–16.

OBORNE, D. J. (1981) *Ergonomics at Work*, Chichester: Wiley.

OBORNE, D. J. and HEATH, T. O. (1979) The role of social space requirements in ergonomics, *Applied Ergonomics*, **10**, 99–103.

OSBORNE, D. J. and LEVIS, J. A. (Eds) (1980) *Human Factors in Transport Research*, London: Academic Press.

ONISHI, N., SAKAI, K. and KOGI, K. (1982) Arm and shoulder muscles load in various keyboard operating postures in women, *Journal of Human Ergology*, **11**, 89–97.

OPCS (1981) *Adult Heights and Weights Survey*, OPCS Monitor ref. 5581/1, London: Office of Population Census and Surveys.

OXENBURGH, M. (1984) Musculoskeletal injuries occurring in world processor operators, *Proceedings of the 21st Annual Conference of the Ergonomics Society of Australia and New Zealand*, 137–43.

PANOFSKY, E. (1970) *Meaning in the Visual Arts*, Harmondsworth: Penguin.

PEPERMANS, R. G. and CORLETT, E. N. (1983) Cross-modality matching as a subjective assessment technique, *Applied Ergonomics*, **14**, 169–76.

PHEASANT, S. T. (1982a) A technique for estimating anthropometric data from the parameters of the distribution of stature, *Ergonomics*, **25**, 981–92.

PHEASANT, S. T. (1982b) Anthropometric estimates for British civilian adults, *Ergonomics*, **25**, 993–1001.

PHEASANT, S. T. (1983) Sex differences in strength: some observations on their variability, *Applied Ergonomics*, **14**, 205–11.

PHEASANT, S. T. (1984a) Human proportions: sex, age and ethnic differences, in E. D. MEGAW (Ed.), pp. 142–7, *Contemporary Ergonomics 1984*, London: Taylor & Francis.

PHEASANT, S. T. (1984b) *Anthropometries: An Introduction for Schools and Colleges*, BSI Educational Publication 7310, London: British Standards Institution.

PHEASANT, S. T. (1987) *Ergonomics: Standards and Guidelines for Designers*, PP 7317, London: British Standards Institution.

PHEASANT, S. T. (1988a) The Zeebrugge–Harrisburg Syndrome, *New Scientist*, 21 January, 55–8.

PHEASANT, S. T. (1988b) User-centred design, in A. S. NICHOLSON and J. E. RIDD (Eds), *Health, Safety and Ergonomics*, London: Butterworths.

PHEASANT, S. T. (1991a) *Ergonomics, Work and Health*, London: Macmillan.

PHEASANT, S. T. (1991b) *Anthropometrics: An Introduction*, 2 Edn, PP 7310, London: British Standards Institution.

PHEASANT, S. T. (1992) Does RSI exist?, *Occupational Medicine*, **42**, 167–8.

PHEASANT, S. T. (1994) Musculoskeletal injury at work: natural history and risk factors, in B. RICHARDSON and A. EASTLAKE (Eds), pp. 146–70, *Physiotherapy in Occupational Health*, Oxford: Butterworth Heinemann.

PHEASANT, S. T. (1994) Repetitive strain injury: towards a clarification of the points at issue, *Journal of Personal Injury Litigation*, September, 223–30.

PHEASANT, S. T., GRIEVE, D. W., RUBIN, T. and THOMSON, S. J. (1982) Vector representations of human strength in whole body exertion, *Applied Ergonomics*, **13**, 139–44.

PHEASANT, S. T. and HARRIS, C. M-T. (1982) Human strength in the operation of tractor pedals, *Ergonomics*, **25**, 53–63.

PHEASANT, S. T. and O'NEILL, D. (1975) Performance in gripping and turning, *Applied Ergonomics*, **6**, 205–8.

PHEASANT, S. T. and SCRIVEN, J. G. (1983) Sex differences in strength: some implications for the design of hand tools, in K. COOMBES (Ed.), pp. 9–13, *Proceedings of the Ergonomics Society's Conference 1983*, London: Taylor & Francis.

PHEASANT, S. T. and STUBBS, D. (1992a) Back pain in nurses: epidemiology and risk assessment, *Applied Ergonomics*, **23**, 226–32.

PHEASANT, S. T. and STUBBS, D. (1992b) *Lifting and Handling: An Ergonomic Approach*, London: National Back Pain Association.

PLAGENHOF, S. (1971) *Patterns of Human Motion: A Cinematographic Analysis*, Englewood Cliffs, NJ: Prentice Hall.

QUINTNER, J. (1991) The RSI syndrome in historical perspective, *International Disability Studies*, **13**, 99–104.

RADL, G. W. (1980) Experimental investigations for optimal presentation mode and colour of symbols on the CRT-screen, in Grandjean and Vigliani (1980), pp. 271–6.

RAMAPRAKASH, D. (Ed.) (1984) *Social Trends*, London: HMSO.

RAMAZZINI, B. (1713) *De Morbis Artificum*, trans. W. C. WRIGHT (1940), *Diseases of Workers*, Chicago: University of Chicago Press.

RAY, R. D. and RAY, W. D. (1979) An analysis of domestic cooker control design, *Ergonomics*, **22**, 124–55.

REBIFFE, R., QUILLIEN, J. and PASQUET, P. (1983) *Enquête anthropométrique sur les conducteurs françaises*, Laboratoire de Physiologie et de Bioméchanique de l'Association Peugeout-Renault.

REBIFFE, R., ZAYANA, O. and TARRIERE, C. (1969) Détermination des zones optimales pour l'emplacement des commandes manuelles dans l'espace de travail, *Ergonomics*, **12**, 913–24.

REYNOLDS, H. M. (1978) The inertial properties of the body and its segments, in NASA (1978), Vol. I, ch. IV.

RIDD, J. E. (1985) Spatial restraints and intra-abdominal pressure, *Ergonomics*, **28**, 149–66.

ROBERTS, D. F. (1973) *Climate and Human Variability: An Addison-Wesley Module in Anthropology*, No. 34, Reading, MA: Addison-Wesley.

ROBERTS, D. F. (1975) Population differences in dimensions, their genetic basis and their relevance to practical problems of design, in A. CHAPANIS (Ed.), *Ethnic Variables in Human Factors Engineering*, Baltimore: Johns Hopkins University Press.

ROCHE, A. F. (1979) Secular trends in stature, weight and maturation, *Monographs of the Society for Research in Child Development*, Serial no. 179; 44 (3–4), 3–27.

ROCHE, A. F and DAVILA, G. H. (1972) Late adolescent growth in stature, *Pediatrics*, **50**, 874–80.

ROEBUCK, JR, J. A., KROEMER, K. H. E. and THOMSON, W. G. (1975) *Engineering Anthropometry Methods*, New York: Wiley.

ROEBUCK, JR, J. A. and LEVENDAHL, B. N. (1961) Aircraft ground emergency exit design considerations, *Human Factors*, **3**, 174–209.

RONA, R. J. (1981) Genetic and environmental factors in the control of growth in childhood, *British Medical Bulletin*, **37**(3), 265–72.

RONA, R. J. and ALTMAN, D. G. (1977) National study of health and growth: standards of attained height, weight and triceps skinfold in English children 5 to 11 years old, *Annals of Human Biology*, **4**, 501–23.

RONA, R. J., SWAN, A. V. and ALTMAN, D. G. (1978) Social factors and health of primary schoolchildren in England and Scotland, *Journal of Epidemiology and Community Health*, **32**, 147–54.

RYAN, G. A. and BAMPTON, M. (1988) Comparison of data processing operators with and without upper limb symptoms, *Community Health Studies*, **12**, 63–8.

SAVINAR, J. (1975) The effect of ceiling height on personal space, *Man–Environment Systems*, **5**, 321–4.

SCHMIDTKE, H. (1980) Ergonomic design principles of alphanumeric displays, in Grandjean and Vigliani (1980) pp. 265–70.

SEARLE, J. A., HARDY, R. N. and HASLEGRAVE, C. M. (1980) A geometrical model for the representation of seat-belt fitting problems, *Ergonomics*, **23**, 305–16.

SHAPIRO, H. (1939) *Migration and Environment*, London: Oxford University Press.

SHOTTON, M. A. (1984) Problems experienced by rheumatism sufferers in the use of standard seat belts, in E. D. MEGAW (Ed.) *Contemporary Ergonomics 1984*, London: Taylor & Francis, pp. 161–6.

SIGLER, P. A. (1943) Relative slipperiness of floor and deck surfaces, National Bureau of Standards Report BM 5100 Washington DC. Cited in Grandjean (1973), p. 284.

SILVERSTEIN, S. J., FINE, L. J. and ARMSTRONG, T. J. (1986) Hand–wrist cumulative trauma disorders in industry, *British Journal of Industrial Medicine*, **43**, 779–84.

SILVERSTEIN, S. J., FINE, L. J. and ARMSTRONG, T. J. (1987) Occupational factors and carpal tunnel syndrome, *American Journal of Industrial Medicine*, **11**, 343–58.

SIMMONDS, G. (1979) Ergonomics standards for road vehicles, *Ergonomics*, **22**, 135–44.

SINGLETON, W. T. (1963) *The Industrial Use of Ergonomics*, Ergonomics for Industry: I, Department of Scientific and Industrial Research.

SMIDT, G. L. (1973) A biomechanical analysis of knee flexion and extension, *Journal of Biomechanics*, **6**, 79–92.

SNOOK, S. H., CAMPANELLI, R. H. and HART, H. N. (1978) A study of three preventive approaches to low back injury, *Journal of Occupational Medicine*, **20**, 478–81.

SNYDER, R. G., SCHNEIDER, L. W., OWINGS, C. L., REYNOLDS, H. M., GOLOMB, D. H. and SCHORK, M. A. (1977) Anthropometry of infants, children and youths to age 18 for product safety design, Report No. DB-270 277, Consumer Product Safety Committee, US Department of Commerce, Bethesda, MD.

SOCIETY OF AUTOMOTIVE ENGINEERS (1974) *Devices for Use in Defining and Measuring Vehicle Seating Accommodation*, SAE J8266, New York: Society of Automotive Engineers.

SOMMER, R. (1969) *Personal Space: The Behavioural Basis of Design*, Englewood Cliffs, NJ: Prentice Hall.

SPENCE, S. (1990) Psychopathology amongst acute and chronic patients with occupationally related upper limb pain versus accident injuries of the upper limbs, *Australian Psychologist*, **25**, 293–305.

SQUIRES, P. C. (1956) The shape of the normal working area, Report No. 275, US Navy Department, New London, CT.

STEIDL, R. E. and BRATTON, E. C. (1968) *Work in the Home*, New York: Wiley.

STOUDT, H. W., DAMON, A. and MCFARLAND, R. (1965) Weight, height and selected body dimensions of adults, National Centre for Health Statistics, Series 11, No. 8.

STOUDT, H. W., DAMON, A. and MCFARLAND, R. A. (1970) Skinfolds, body girths, biacromial diameter and selected anthropometric indices of adults, National Centre for Health Statistics, Series 11, No. 35.

TANNER, J. M. (1962) *Growth at Adolescence*, Oxford: Blackwell.

TANNER, J. M. (1978) *Foetus into Man*, London: Open Books.

TANNER, J. M., HAYASHI, T., PREECE, M. A. and CAMERON, N. (1982) Increase in length of leg relative to trunk in Japanese children and adults from 1957–1977: a comparison with British and with Japanese Americans, *Annals of Human Biology*, **9**, 411–23.

TANNER, J. M. and WHITEHOUSE, R. H. (1976) Clinical longitudinal standards for height, weight, height velocity, weight velocity and stages of puberty, *Archives of Diseases in Childhood*, **51**, 170–9.

TANNER, J. M., WHITEHOUSE, R. H. and TAKAISHI, M. (1966) Standards from birth to maturity for height, weight, height velocity and weight velocity: British children, 1965, Part I, *Archives of Diseases of Childhood*, **41**, 454–71; Part 2, *Archives of Diseases of Childhood*, **41**, 613–35.

TAYLOR, J. H. (1973) Vision, in J. T. PARKER and V. R. WEST (Eds), *Bioastronautics Data Book*, Washington, DC: NASA.

THOMPSON, A. R., PLEWES, L. W. and SHAW, E. G. (1951) Peritendinitis and simple tenosynovitis: a clinical study of 544 cases in industry, *British Journal of Industrial Medicine* **8**, 150–60.

THOMPSON, D. and BOOTH, R. T. (1982) The collection and application of anthropometric data for domestic and industrial standards, in R. EASTERBY, K. H. C. KROEMER and D. B. CHAFFIN (Eds), *Anthropometry and Biomechanics: Theory and Applications,* New York: Plenum.

THOMSON, A. M. (1959) Maternal stature and reproductive efficiency, *Eugenics Review,* **51**, 157–62.

TICHAUER, E. R. (1978) *The Biomechanical Basis of Ergonomics,* New York: Wiley.

TILDESLEY, M. F. (1950) The relative usefulness of various characters on the living for racial comparison, *Man,* **50**, 14–18.

TRAVELL, J. (1967) Mechanical headache, *Headache,* **7**, 23–9.

TRAVELL, J. E. and SIMONS, D. E. (1983) *Myofascial Pain and Dysfunction: The Trigger Point Manual,* Baltimore: Williams & Wilkins.

TROTTER, M. and GLESER, G. (1951) The effect of ageing upon stature, *American Journal of Physical Anthropology,* **9**, 311–24.

TROUP, J. D. G. and EDWARDS, F. C. (1985) *Manual Handling: A Review Paper,* London: HMSO.

TUTT, D. and ADLER, D. (Eds) (1979) *New Metric Handbook,* London: Architectural Press.

VAN COTT, H. P. and KINKADE, R. G. (Eds) (1972) *Human Engineering Guide to Equipment Design,* Washington, DC: US Department of Defense.

VAN WELY, P. (1970) Design and disease, *Applied Ergonomics,* I, 262–9.

VIHMA, T., NORMINEN, M. and MUTANEN, P. (1982) Sewing machine operators' work and musculo-skeletal complaints, *Ergonomics,* **25**, 295–8.

VIITISALO, J. T., ERA, P., LESKINEN, A. L. and HEIKKINEN, E. (1985) Muscular strength profiles and anthropometry in random samples of men aged 31–35, 51–55 and 71–75 years, *Ergonomics,* **28**, 1563–74.

VINCENT, L. M. (1979) *Competing with the Sylph: Dancers in Pursuit of the Ideal Body Form,* Kansas City: Andrews & McMeel.

WARD, J. S. (1971) Ergonomic techniques in the determination of optimum work surface heights, *Applied Ergonomics,* **2**, 171–7.

WARD, J. S. and BEADLING, W. M. (1970) Optimum dimensions for domestic staircases, *Architects Journal,* **151**, 513–20.

WARD, J. S. and KIRK, N. S. (1970) The relation between some anthropometric dimensions and preferred working surface heights in the kitchen, *Ergonomics,* **6**, 783–97.

WARIS, P. (1979) Occupational cervico-brachial syndromes: a review, *Scandinavian Journal of Work Environment and Health,* **5**, Supplement 3, 3–14.

WATERS, T. R., PUTZ-ANDERSON, V., GARG, A. and FINE, L. J. (1993) Revised NIOSH equation for the design and evaluation of manual lifting tasks, *Ergonomics,* **36**, 749–76.

WEINER, J. S. (1982) The measurement of human workload, *Ergonomics,* **25**, 953–66.

WELLS, L. H. (1963) Stature in earlier races of mankind, in D. BOTHWELL and E. HIGGS (Eds), *Science in Archaeology,* London: Thames & Hudson.

WESTGAARD, R. H. and AARAS, A. (1980) Static muscle load and illness among workers doing electro-mechanical assembly work, Institute of Work Physiology, Oslo.

WESTON, H. C. (1953) Visual fatigue with special reference to lighting, in W. F. FLOYD and A. T. WELFORD (Eds), *Symposium on Fatigue,* London: H. K. Lewis.

WHO (1980) *International Classification of Impairments, Disabilities and Handicaps,* Geneva: World Health Organization.

WICKSTROM, P. (1979) Effect of work on degenerative back disease, *Scandinavian Journal of Work Environment and Health,* **4**, Supplement 51, 1–12.

WIGLEY, R. D. (1990) Repetitive strain syndrome: fact not fiction, *New Zealand Medical Journal,* 28 February, 75–6.

WILLIAMS, J. C. (1977) Passenger-accompanied luggage, *Applied Ergonomics,* **8**, 151–7.

WILMORE, J. H. (1976) *Athletic Training and Physical Fitness: Physiological Principles and Practices of the Conditioning Process,* Boston: Allyn & Bacon.

WILSON, J. R. and CORLETT, E. N. (1995) *Evaluation of Human Work*, London: Taylor & Francis.

WILSON, J. R., COOPER, S. E. and WARD, J. S. (1980) *A Manual of Domestic Activity Space Requirements*, Loughborough: Institute for Consumer Ergonomics.

WINTER, D. A. (1979) *Biomechanics of Human Movement*, New York: Wiley.

WISNER, A. and REBIFFE, R. (1963) Methods of improving workspace layout, *International Journal of Production Research*, **2**, 145–67.

WOOD, P. H. N. (1975) *Classification of Impairments and Handicaps*, Geneva: World Health Organization.

WOODSON, W. E. (1981) *Human Factors Design Handbook*, New York: McGraw-Hill.

WOODSON, W. E. and CONOVER, D. W. (1964) *Human Engineering Guide for Equipment Designers*, Berkeley: University of California Press.

WRIGHT, P. and BARNARD, P. (1975) Just fill in this form: a review for designers, *Applied Ergonomics*, **6**, 213–20.

YAMADA, H. (1970) *Strength of Biological Materials*, Baltimore: Williams & Watkins.

YAMANA, N., OKABE, K., NANAKO, C., ZENITANI, Y. and SAITA, J. (1984) The body form of pregnant women in monthly transitions, *Japanese Journal of Ergonomics*, **2**, 171–8 (in Japanese).

YANAGISAWA, S. and KONDO, S. (1973) Modernization of physical features of the Japanese with special reference to leg length and head form, *Journal of Human Ergology*, **2**, 97–108.

Index

flexion 57, 58
of spine and hips 70–1, 73, 75
of wrist and arms 83, 85
flooring materials 121
foot breadth and length
body dimensions for 42
data for 178–213
footrests 94, 95
forward-tilt seating 73–5
France 175, 186
frequency distributions 16–21, 215–23
functionalism 8, 9, 10

Gaussian distributions *see* normal distributions
gender 154–8, 161–4, 170–1
genetic factors, in secular change 167–8
Georgian chairs 9, 10
girls *see* children and teenagers
golfer's elbow 139
grip reaches
body dimensions for 44
data for 178–93, 197–213
gripping 84, 86–90, 143–4
growth, human 160–8
guards for machinery 121, 160
guidelines *see* standards and guidelines

hand breadth and length
body dimensions for 41–2
data for 178–213
handbasins 111
handles 86–92, 132
see also hands
handling loads *see* lifting and handling
hands 83–6
see also arms; handles; work-related upper
limb disorders; wrists
hazard 117–18
head, posture of 62, 63–5, 100
head breadth and length
body dimensions for 36, 37, 40
data for 178–213
Health and Safety Executive 97, 135–7
Health and Safety at Work Act 118, 122
height *see* stature; working height
Herald of Free Enterprise 120–1
hip breadth
body dimensions for 38
data for 178–213
sex differences in 155–6
hip height
body dimensions for 32
data for 178–93, 197–213
historical aspects 4, 7–8, 9–10
see also secular trend
Holland 174–5, 176, 185
homes 105–14
Hong Kong 176, 192
horizontal reference plane 30
HSE 97, 135–7
human proportion *see* proportion, human
human variability 16–21, 153–73

India 165, 175, 191
industrial work

bench height for 25–6, 65, 145
work-related upper limb disorders 92, 137,
139, 140, 141, 142–4
infants
anthropometric data 176, 194–6
growth and development 160–1, 163
measurements 30–1
injuries and diseases
accidents at work 115–22, 123
back injuries *see* back disorders
ergonomic type defined 122–3
eyestrain 64, 65
keyboard 3–4, 137–8, 140, 145–50
lifting and handling 116, 126–37
over-exertion injuries 123, 134
over-use injuries 123, 133, 138–50
upper limb disorders *see* work-related upper
limb disorders

Japan 159, 160, 165, 176, 193
joints
flexibility 57–9, 169–70
and posture 63, 66–7

keyboard work 93–4, 98–104
furniture for 65–6, 94, 98
injuries from 3–4, 137–8, 140, 145–50
see also office work; screens
kinetic chain 84, 86
kitchens 65, 105–9
knee height 34, 36, 178–213
kneeling chairs 74–5
kneeling height and leg length 47
knuckle height
body dimensions for 33
data for 178–93, 197–213
as optimum lifting height 129–30
kyphosis 70

laptop computers 100
lateral epicondylitis 139
layout
of kitchens 105–6
of workspace 46, 128–32
lectern, fitting trial for 23–5
leg room 78–9
legal aspects
of cost–benefit trade-offs 6
following *Herald of Free Enterprise* 120
of lifting injuries 134–5
of safety at work 118–19, 122
of upper limb disorders 3–4, 140
Leonardo da Vinci 7–8
leptokurtic distributions 19
lifting and handling 116, 126–8
weight limits 67, 132–7
workspace layout 65, 128–32
lifting zones 131, 134–6
lighting, for screen-based work 98
limiting user, principle of 14, 23, 122
limits
maximum permissible (lifting) 134–5
method of 25–8
lordosis 69–70, 71–2, 73
lumbar region 69